MULTILATERAL DEVELOPMENT BANK SUPPORT FOR DISASTER-RESILIENT INFRASTRUCTURE SYSTEMS

AUGUST 2022

ASIAN DEVELOPMENT BANK

MULTILATERAL DEVELOPMENT BANK SUPPORT FOR DISASTER-RESILIENT INFRASTRUCTURE SYSTEMS

AUGUST 2022

ADB

ASIAN DEVELOPMENT BANK

Contents

Tables, Figures, and Boxes

TABLES

FIGURES

BOXES

Acknowledgments

This publication was prepared under the Asian Development Bank (ADB) technical assistance project Building Disaster-Resilient Infrastructure through Enhanced Knowledge. Grant funding for the project came from the Japan Fund for Prosperous and Resilient Asia and the Pacific (JFPR), financed by the Government of Japan through ADB.

Steven Goldfinch, senior disaster risk management specialist with ADB's Sustainable Development and Climate Change Department (SDCC), provided overall guidance. Mario Unterwainig, former SDCC disaster risk management specialist (Resilient Infrastructure), led the development of the publication with Alih Faisal Pimentel Abdul, coordinating consultant. Ghia V. Rabanal, operations analyst, and Gren J. Saldevar, senior operations assistant, both with SDCC, provided administrative support. The publication was edited by Mary Ann Asico, and layout was done by Rocilyn Locsin Laccay. Vivid Economics contributed to the research, gathered data, and prepared the draft report.

The publication benefited from peer review and detailed comments from Esmyra P. Javier, climate change specialist (Climate Finance) with ADB's SDCC. Detailed interviews were held with the following:

- **Asian Development Bank.** Charlotte Benson, SDCC principal disaster risk management specialist; and Arghya Sinha Roy, SDCC principal climate change specialist (climate change adaptation).
- **African Development Bank**. Dieudonné Goudou, principal climate risk and disaster officer with the ClimDev Special Fund; James Kinyangi, chief, Climate and Policy; Mike Salawou, division manager, Infrastructure & Partnerships; and Donald Singue, climate and disaster risk finance specialist.
- **Asian Infrastructure Investment Bank**. Xing Zhang, senior climate policy specialist.
- **European Bank for Reconstruction and Development.** Zeynep Cansever, monitoring, reporting, and verification associate; and Sarah Duff, principal, Green Economy and Climate Action.
- **European Investment Bank.** Pieter Coppens, senior economist; Cinzia Losenno, senior climate change specialist, Environment, Climate, and Social Office; Michael Rattinger, senior officer, Operations Department; and Roman Roehrl, climate change specialist, Environment, Climate, and Social Office.
- **Inter-American Development Bank.** Maricarmen Esquivel Gallegos, climate change senior specialist; Sergio Lacambra Ayuso, coordinator, Disaster Management Cluster, and sector principal, Natural Disaster and Risk Management; and Maria Alejandra Escovar Bernal, disaster and climate change risk consultant.
- **Islamic Development Bank.** Moustapha Diallo, disaster risk management specialist; and Olatunji Yusuf, senior climate change specialist, Resilience and Social Development Global Practice.
- **World Bank.** Frederico Pedroso, disaster risk management specialist.

ADB thanks everyone who helped in the development of this publication.

Abbreviations

ADB	Asian Development Bank
AfDB	African Development Bank
AIIB	Asian Infrastructure Investment Bank
DMC	developing member country
DRF	disaster risk financing
EBRD	European Bank for Reconstruction and Development
EIB	European Investment Bank
GFDRR	Global Facility for Disaster Reduction and Recovery
IDB	Inter-American Development Bank (alternative abbreviations: "IADB," "IaDB")
IsDB	Islamic Development Bank
MDB	multilateral development bank
NBS	nature-based solution

Executive Summary

The disaster resilience of infrastructure systems is a critical challenge for developing Asia. Exposure to climate and geophysical hazards is already widespread. Between 2004 and 2020, disasters led to losses of over $500 billion in the region, affecting 2.1 billion people (Sirivunnabood and Alwarritzi 2020). The risk posed by natural hazards is expected to increase notably in the coming decades as economies grow, urbanize, and come to grips with climate change. Infrastructure has a central role to play in supporting economic resilience. Large-scale spending on infrastructure will underpin economic development, and the way in which the systems are planned, operated, and financed will fundamentally determine how well the region can withstand, and recover from, natural calamities. To sustain economic development and reduce poverty, the region must have disaster-resilient infrastructure systems, with provisions for reducing, transferring, and managing the climate and disaster risks to the systems.

Multilateral development banks (MDBs) have both the capacity and the aspiration to play a crucial role in promoting disaster-resilient infrastructure in the coming decades. Their direct contribution to infrastructure investments in the developing member countries (DMCs) of the Asian Development Bank (ADB) in 2015 amounted to 2.5% of the total—10% excluding the People's Republic of China and India (ADB 2017d). MDBs can likewise have a key role in crowding in private sector investments in infrastructure, to meet significant infrastructure needs across developing Asia and globally. Climate and disaster resilience is one of the seven operational priorities identified in ADB's Strategy 2030 (ADB 2018b), and a common focus shared by other MDBs.

This report identifies opportunities for MDBs to provide effective support to their members in improving infrastructure resilience. The opportunities are specifically aimed at addressing the challenges faced by members despite the MDBs' current offerings, and are presented here together with best-practice examples showcasing successes achieved by MDBs worldwide. The discussion draws on an extensive review of the literature, as well as on original insights from stakeholders in ADB's DMCs and from numerous experts in eight MDBs. This report highlights 11 opportunities across the following three main themes based on these insights.

First, infrastructure resilience can be enhanced through consistent MDB involvement throughout the infrastructure life cycle, starting with early planning and extending beyond the completion of project-specific support. MDB support is currently sought only after important infrastructure design choices have been made, and is often limited to the duration of a specific phase of the project (e.g., asset construction). Early involvement of MDBs in project planning, their continued support throughout the asset life cycle, and embedding of project-specific initiatives in the strategic development support they provide can magnify MDB impact in promoting infrastructure resilience (Opportunities 1, 2, and 7).

Second, MDBs can build a strong business case for resilience by improving the understanding and accessibility of user-friendly risk data and targeted finance products. Disaster risks and resilience benefits continue to be underestimated in investment decisions. Decision-makers are often unaware of the full range of benefits infrastructure resilience can provide, or lack the data or the technical capacity needed to assess them. MDBs, with their considerable expertise in disaster risk and resilience assessments, can play a major role in making user-friendly risk information more readily accessible. Once the risks and benefits are understood, these can be managed effectively by public and private stakeholders within their projects, with the help of targeted infrastructure and disaster risk finance solutions from MDBs (Opportunities 3, 5, 6, 8, and 9).

Third, increased collaboration and knowledge sharing within and across MDBs and their members can accelerate progress. Knowledge sharing, according to stakeholders, is often informal and ad hoc. More regular, structured exchanges between different teams and organizations could significantly improve the uptake of new insights and make it more consistent and timely. This process is particularly important for emerging topics of shared interest, such as the use of nature-based solutions within infrastructure systems (Opportunities 4, 10, and 11).

If these opportunities are realized by the MDBs, the resulting action plans could be vitally important in bridging the infrastructure resilience gap in developing countries in Asia and the Pacific, and beyond.

1 Introduction

Background

Resilient infrastructure plays a vital role in securing development objectives. This is recognized in Sustainable Development Goal 9: "Build resilient infrastructure, promote inclusive and sustainable industrialization, and foster innovation" (United Nations Department of Economic and Social Affairs [UNDESA] n.d.). It is also part of the third priority area of the Sendai Framework for Disaster Risk Reduction: "Investing in disaster risk reduction for resilience" (United Nations Office for Disaster Risk Reduction [UNDRR] 2015). Infrastructure development has helped unlock the rapid economic growth and poverty reduction seen over the past decades across Asia and the Pacific, as well as in developing countries elsewhere.

However, economic growth and the infrastructure systems that underpin it are imperiled by increasing disaster risk. The frequency and severity of disasters is projected to increase worldwide, making the disaster resilience of infrastructure a key global concern (Intergovernmental Panel on Climate Change [IPCC] 2021; Milman et al. 2021). Many developing countries, including several in Asia and the Pacific, are exposed to disproportionately high levels of risk from both geophysical and climate hazards. Disaster risks and their interaction with infrastructure systems vary, but much of the region is exposed to severe risk from at least one hazard. Many countries encompass multihazard hot spots, which pose the largest and most complex risk (Lu 2019). These risks are set to increase further over the coming years and decades as both climate and population patterns change. According to the Sixth Assessment Report of the IPCC, it is very likely that heavy precipitation events will intensify and become more frequent under added global warming (IPCC 2021). Similarly, other hazards, including tropical cyclones, coastal flooding, heat waves, and droughts, are expected to increase in frequency and severity. In addition to this greater risk of acute disasters, climate change is projected to cause significant long-term stresses from chronic hazards, such as rising sea levels and higher ocean temperatures. Sea level rise is particularly relevant to many low-lying islands in the Pacific, as well as to Asian coastal regions. At the same time, insufficiently risk-informed development, unplanned urbanization, and population growth will increase the number of people and the value of economic activity in hazard-exposed areas, further heightening the adverse impact of disasters. For example, direct infrastructure asset damage from a 1-in-100-year flood event in Ho Chi Minh City could rise from about $200–$300 million today to $0.5–$1 billion in 2050, with the estimated costs of knock-on effects to the economy rising from $100–$400 million to $1.5–$8.5 billion (Woetzel et al. 2020).

High infrastructure investment needs offer a unique opportunity to improve resilience. Infrastructure investment needs across developing member countries (DMCs) of the Asian Development Bank (ADB) alone are estimated at $26 trillion for 2016–2030 (ADB 2017d). This unparalleled magnitude of investment in infrastructure offers an equally unprecedented opportunity to make infrastructure systems more resilient. Conversely, ill-designed infrastructure investments over the coming decade can lock in long-term vulnerabilities, and risk eroding developmental gains.

Multilateral development banks (MDBs) have both the capacity and the aspiration to play a crucial role in promoting disaster-resilient infrastructure in the coming decades. Their direct contribution to infrastructure investments in ADB DMCs in 2015 amounted to 2.5% of the total—10% excluding the People's Republic of China and India (ADB 2017d). MDBs can also play a key role in crowding in private sector investments in infrastructure, to meet significant infrastructure needs across developing Asia and globally. Climate and disaster resilience is one of the seven operational priorities identified in ADB's Strategy 2030 (ADB 2018b), and a common focus shared by several other MDBs. For example, the third pillar of the Disaster Risk and Resilience Strategy of the Islamic Development Bank (IsDB) is investing in resilient infrastructure and disaster risk information, and IsDB intends to continue developing tools for managing climate and disaster risks in its member states (IsDB 2020a, 2020d).[1] Integrating disaster risk management into economic development plans is also a stated core sector of work of the Inter-American Development Bank (IDB) (IDB, n.d.[e]).[2] Support for disaster-resilient infrastructure is therefore a central theme of MDBs' mandates.

Objective and Approach

This report was prepared to highlight relevant opportunities for MDBs to support their DMCs in achieving more resilient infrastructure assets, networks, and systems. These opportunities, brought out in the following two subsections in Sections 2, 3, and 4 of this report, will enable MDBs to draft an infrastructure resilience strategy suited to their priorities, requirements, and circumstances:

- **Current multilateral development bank offerings.** First, the current MDB offerings are analyzed. MDBs can thus learn from existing resilience initiatives of their peers in other regions in the world. The assessment can also spur inter-MDB dialogue on "best practices."

- **Opportunities for future multilateral development bank offerings.** Second, opportunities for MDBs to support developing countries in overcoming barriers to resilience improvements are presented. MDBs can identify the most relevant challenges for their DMCs, given their regional context, and prioritize the opportunities that address those challenges.

If these opportunities are realized by MDBs, the resulting action plans could play a crucial role in bridging the infrastructure resilience gap in developing countries.

This report draws on deep insights developed through interviews with MDB disaster risk management and infrastructure experts. Interviews were conducted with experts across eight MDBs—ADB, the African Development Bank (AfDB), the Asian Infrastructure Investment Bank (AIIB), the European Bank for Reconstruction and Development (EBRD), the European Investment Bank (EIB), IDB, IsDB, and the World Bank Group. The interviews had both directive and nondirective components. First, all experts were asked identical questions about the existing disaster risk management (DRM) structure, key infrastructure resilience initiatives, and areas of expansion or improvement pertaining to their organization. The second part of the interview was a deep dive into each MDB's specific approaches, instruments, and initiatives, guided by research gaps identified during the literature review that was conducted concurrently.

Interview findings were complemented with a systematic review of recent MDB publications laying out current offerings and approaches to supporting infrastructure resilience in developing countries. Besides the interviews with experts, this report was informed by a desk-based review of project-level and institutional support provided by MDBs to promote effective resilience planning, implementation, risk management, and financing in developing countries. Insights were gathered systematically from the websites and publications of the eight MDBs mentioned above. The review covered more than 220 documents, including MDB disaster risk management policies and strategies, climate action plans, annual reports, project appraisal policies, and joint MDB reports on climate finance.

---

[1]   IsDB. 2019. *Disaster Risk Management and Resilience Policy*.
       IsDB. 2020. *2020-2025 Climate Action Plan*.
[2]   IaDB. Environment and Natural Disasters.

Report Structure

This report is structured around three core areas in which MDBs can promote the resilience of their clients' infrastructure systems, as shown in Figure 1:

- risk-informed infrastructure planning, through disaster risk assessments, to identify resilience priorities and inform infrastructure strategies and investments;
- financing assistance, to make the clients' infrastructure systems more resilient by enabling improvements in infrastructure resilience before a disaster occurs, swift response, and recovery ("building back better") after a disaster has occurred; and
- knowledge building through regional and global networks, connecting practitioners and scaling up effective mechanisms.

Figure 1: Areas of Multilateral Development Bank Support for Infrastructure Resilience

Knowledge building
Enable long-term capacity building, knowledge exchange, and good practice

Risk-informed planning
Identify and assess disaster risks; integrate findings into planning processes

Financing mechanisms
Support disaster-resilient infrastructure through infrastructure and disaster risk finance

Source: Vivid Economics

For each area, the report summarizes key mechanisms already employed by MDBs, and then discusses opportunities to further evolve or scale up current offerings to best meet DMC needs. Challenges and needs faced by developing countries highlighted in the report are based on a review of existing progress and barriers to disaster-resilient infrastructure in ADB's DMCs, which are presented in greater detail in a separate, complementary publication (ADB 2022). Common challenges include a lack of capacity and data to assess disaster risks, an incomplete view of resilience benefits in the cost–benefit analysis underlying investment decisions, and insufficient alignment and coordination of stakeholders.

2   Risk-Informed Planning

This section reviews current MDB activity and future opportunities to promote risk-informed planning. To encourage resilient infrastructure development, decision-makers must be helped to understand how disaster risks affect, and are affected by, infrastructure systems. A risk-informed planning process can provide this understanding. Therefore, this section considers

- project-level risk assessment and planning, which seeks to ensure that infrastructure projects or programs financed by MDBs adhere to disaster resilience standards, as well as

- broader strategic support for the development of sector or national infrastructure plans.

Figure 2: Risk-Informed Planning—Challenges and Opportunities

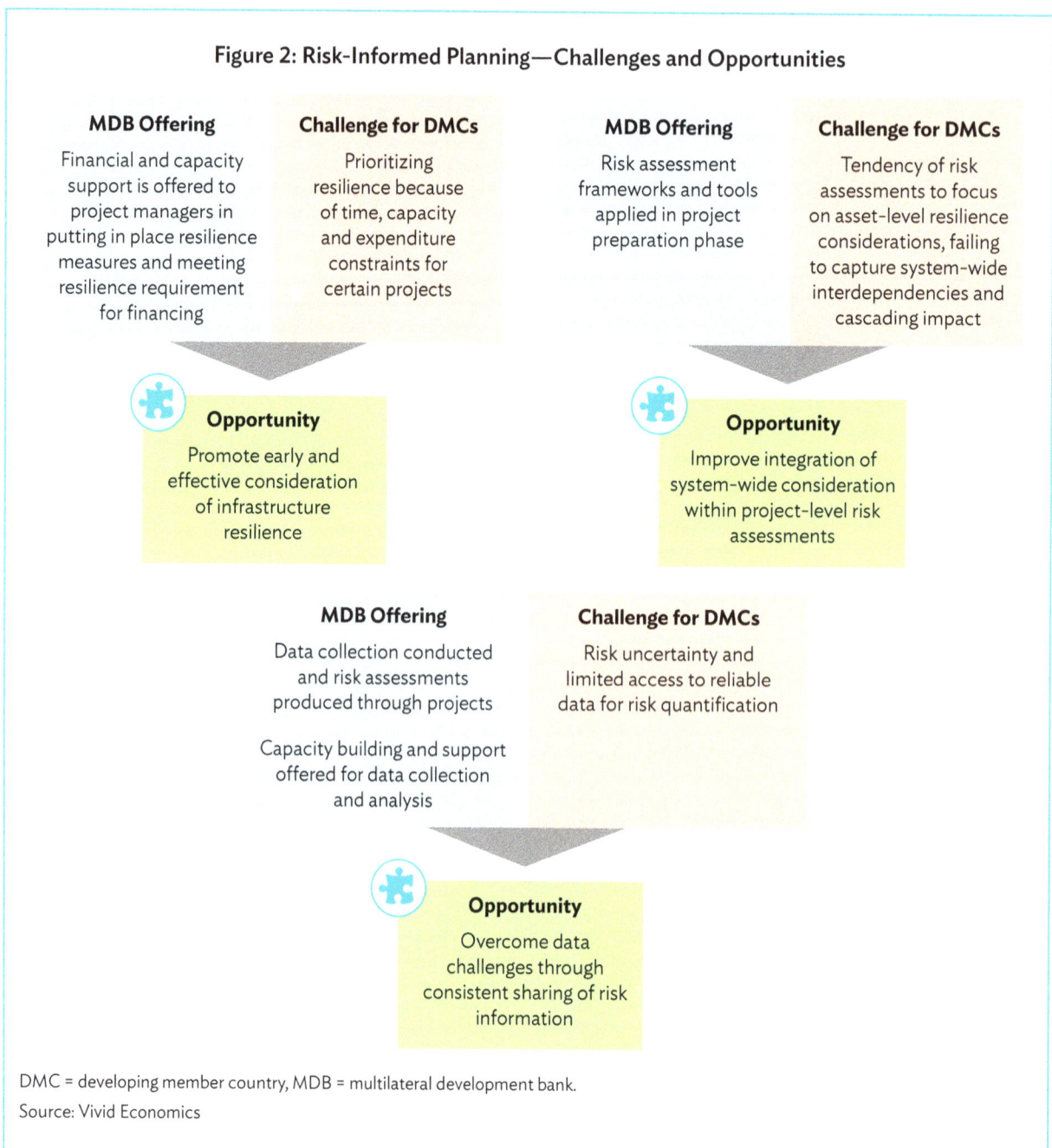

MDB Offering

Financial and capacity support is offered to project managers in putting in place resilience measures and meeting resilience requirement for financing

Challenge for DMCs

Prioritizing resilience because of time, capacity and expenditure constraints for certain projects

Opportunity

Promote early and effective consideration of infrastructure resilience

MDB Offering

Risk assessment frameworks and tools applied in project preparation phase

Challenge for DMCs

Tendency of risk assessments to focus on asset-level resilience considerations, failing to capture system-wide interdependencies and cascading impact

Opportunity

Improve integration of system-wide consideration within project-level risk assessments

MDB Offering

Data collection conducted and risk assessments produced through projects

Capacity building and support offered for data collection and analysis

Challenge for DMCs

Risk uncertainty and limited access to reliable data for risk quantification

Opportunity

Overcome data challenges through consistent sharing of risk information

DMC = developing member country, MDB = multilateral development bank.
Source: Vivid Economics

Current Multilateral Development Bank Offerings

MDBs support risk-informed planning through risk assessments for projects they finance. MDBs generally define resilience requirements for all infrastructure investments they finance, screening infrastructure projects that are funded, to ensure resilience to disaster risks. These requirements are implemented via a risk management framework. Alongside disaster risks, other topics within a broader range, such as displacement or environmental risks induced by the project, are considered through social and environmental safeguards.

Risk management frameworks vary in detail across banks, but generally include a first screening step, followed by more detailed assessments for a subset of projects. Projects are classified as high-, moderate-, or low-risk through screening. After this first stage, high- and moderate-risk projects undergo a more extensive risk assessment, which typically combines both qualitative and quantitative appraisal. For a project to be approved, key risks must be identified, and their impact understood, costed, and mitigated with appropriate measures. Capacity-building initiatives around the use and understanding of risk assessment tools are also deployed at times. More information about the tools currently used by MDBs to conduct these screenings can be found in the table.

Table 1: Risk Screening and Assessment Tools Used by Multilateral Development Banks

Tool	Description
National-/Policy-Level Tool	This tool designed by the World Bank supports national and sectoral strategies. It provides users—decision-makers or MDBs—with an overall climate and disaster risk score based on an analysis of national priority sectors, predicted climate impact, and institutional readiness. It does not provide a detailed risk analysis but highlights the need for further studies in relevant cases.[a]
Project-level tools	Project-level tools designed by the World Bank cover the agriculture, water, roads, coastal flood protection, energy, health, and general (non-road transportation, community development, education, information and communication technology, natural resources, mining and metals, natural resources) sectors. They can offer rapid or in-depth screening for climate and disaster risk and are available for use by development practitioners. They do not provide detailed risk analysis but highlight the need for further studies in relevant cases (World Bank, n.d.[a], n.d.[l]).[b]
AWARE	AWARE is an online climate risk screening tool used systematically by several MDBs—ADB, AIIB, IADB, IsDB. It delivers a risk rating for each relevant natural hazard.[c] In some MDBs, a high score will lead to further in-depth disaster risk screening.[d]
Rapid Assessment Tool for Energy and Climate Adaptation (ATECA) Quick View	This tool designed by the World Bank enables the screening of climate risks in the renewable energy sector.[e]
Energy Sector Management Assistance Program (ESMAP) Hands-on Energy Adaptation Toolkit	This qualitative support framework designed by the World Bank allows policy-makers and decision-makers to assess vulnerabilities in the energy sector.[f]
CityStrength Diagnostic	This qualitative tool built by the World Bank and GFDRR assesses cities' adaptation needs and resilience options across all sectors.[g]
Urban Risk Assessment	This methodology provided by the World Bank allows city and project managers to find the most relevant measures for identifying and assessing a city's risk.[e]

Table continued on next page

Confronting Climate Uncertainty in Water Resources planning and Project Design: The Decision Tree Framework	With this qualitative framework designed by the World Bank, project managers in the water sector can evaluate the climate robustness of their projects.[h]
Screening toolkit, with in-house GIS platform	To the IADB applies an in-house screening toolkit, which automatically provides a disaster and climate risk classification for the initiative assessed. This is supported by a GIS platform. Projects deemed at high or medium risk then undergo in-depth qualitative and quantitative risk assessments. The IADB recommends specific tools for conducting these assessments, depending on the project characteristics and sector.[i]
Scorecard system	The AfDB uses a set of scorecards with relevant questions on project characteristics, to assess projects and classify them as low-, medium-. or high-risk. High- and medium-risk projects require further evaluation and assessment of adaptation measures with other tools.[j]
In-house physical climate risk screening tool	This tool, designed and used by the EBRD, allows each project to be screened against different physical hazards. The resulting score determines whether further screening is necessary, according to the bank's procedures.[k]
E&S Risk Management Toolkit	This toolkit, designed and used by the EBRD, allows financial institutions to screen transactions for environmental and social risks.[l]
Climate Risk Assessment tool	This tool was designed, and is used by, the EIB to screen and assess its activities against climate risks. If a project is found to be at high risk, further assessments will be carried out by the bank in accordance with sector-specific methodologies.[m]
Community-based Risk Screening Tool – Adaptations and Livelihoods (CRiSTAL)	This tool is used by several MDBs to evaluate which community-relevant climate risks their projects can address.[n]

ADB = Asian Development Bank, AfDB = African Development Bank, EBRD = European Bank for Reconstruction and Development, EIB = European Investment Bank, GEDRR = Global Facility for Disaster Reduction and Recovery.

[a] World Bank. National/Policy Level Tool: Climate and Disaster Risk Screening Guidance Note (accessed January 2021).
[b] World Bank. Climate & Disaster Risk Screening Tools: About the Tools (accessed January 2022); World Bank. World Bank Climate and Disaster Risk Screening Tools (accessed January 2022).
[c] Environmental XPRT. Acclimatise Aware – Online, Rapid, Climate Risk Screening Tool (accessed January 2022).
[d] ADB. 2017. Disaster Risk Assessment for Project Preparation: A Quick Guide. Manila.
[e] World Bank. Climate & Disaster Risk Screening Tools: Complementary Risk Analysis Tools and Guidance (accessed January 2021).
[f] ESMAP. 2010. HEAT: Hands-On Energy Adaptation Toolkit. Washington, DC: World Bank.
[g] World Bank. 2017. The CityStrength Diagnostic: Promoting Urban Resilience. 17 October.
[h] P. A. Ray and C. M. Brown. 2015. Confronting Climate Uncertainty in Water Resources Planning and Project Design: The Decision Tree Framework. Washington, DC: World Bank.
[i] M. Barandiarán et al. 2019. Disaster and Climate Risk Assessment Methodology for IDB Projects: A Technical Reference Document for IDB Project Teams. Technical Note No. TN-01771. Washington, DC: Inter-American Development Bank.
[j] AfDB. 2022. Climate Safeguards System (CSS): Climate Screening and Adaptation Review & Evaluation Procedures Booklet. Abidjan, Côte d'Ivoire.
[k] EBRD. 2021. Task Force on Climate-Related Financial Disclosures Report 2020. London.
[l] EBRD. Supporting Tools and Toolkit. EBRD Environmental and Social Risk Management Manual (E-Manual) (accessed January 2022).
[m] EIB. 2020. EIB Climate Strategy. Luxembourg..
[n] CRiSTAL. CRiSTALTool.org: Community-based Risk Screening Tool – Adaptation and Livelihoods (accessed January 2022).
Note: This table summarizes the different tools used by MDBs to conduct project risk assessments. For MDBs where details on specific tools were not available, this table describes instead the relevant broader risk assessment procedure.
Source: Vivid Economics, based on references indicated in the table.

In addition, MDBs offer strategic risk assessment and planning support at both the sector-specific and the national level:

- Sector ministries can call on MDB support in conducting sector-wide risk assessments and defining standards, including risk-informed zoning, and ensuring that these standards are maintained. MDBs provide capacity and institutional support for standardized and sector-wide approaches to infrastructure resilience planning. This may include support around incorporating uncertainty about future climate change impact into decision-making.

- MDBs support national governments and cross-sector ministries in assessing and prioritizing risks of national significance. This support is achieved, for example, by making infrastructure resilience a key component of country partnership strategies and ensuring that it is appropriately integrated into long-term national development strategies. Crosscutting, comprehensive support programs can address various aspects of infrastructure resilience in an integrated, system-wide approach. Examples are the City Resilience Program, the Global Facility for Disaster Reduction and Recovery (GFDRR), the AfDB Global Centre on Adaptation, and the AIIB Sustainable Cities Strategy.

In many cases, these strategic initiatives are further underpinned with knowledge products and tools, which are explored in Section 4.

Opportunities for Future Multilateral Development Bank Offerings

Opportunity 1: Promote early and effective consideration of infrastructure resilience

Challenge

Despite the importance of early consideration, developing countries find it challenging to prioritize resilience measures in the project planning stage because of time, capacity, and expenditure constraints. The window for resilience to be incorporated is narrow, and changing paths after initial planning parameters have been set can be costly. For example, if the need for a bridge across a river is identified as part of a national infrastructure plan, there is scope to determine its location in a risk-informed way; if, by contrast, a proposal with a specific location is presented to a financier, the costs of incorporating resilience are significantly higher. Similarly, if core design options—such as material used or height— have already been decided or implemented, less room is left to make the project more resilient to disasters. Despite the clear advantages of considering resilience measures from the outset, developing countries may find this challenging in light of the increasing need to close the infrastructure gap, which imposes time, capacity, and expenditure constraints on the planning process.

Moreover, project-specific planning processes are often not integrated into broader strategic frameworks, making user resilience less achievable on a national scale. The planning processes for individual infrastructure projects are not always systematically linked with national or sector-wide planning. Funding that is project-based and not systematically allocated through larger strategies could be the reason. Some infrastructure owners may also prioritize project-specific planning because of the nature and geography of their activities—if they operate through a localized regulated monopoly, for instance. Moreover, complexities in governance structures can also cause difficulties in integration. Indeed, sector agencies and national governments have a scope and level of planning unlike those of infrastructure owners and operators. The identification of relevant projects and integration within broader pipelines can, in turn, become more difficult. Disjointed planning can lead to missed opportunities to optimize system-wide resilience when project-specific decisions are made.

Multilateral Development Bank Opportunity

MDBs can take a more proactive role in project conception, including contributing to critical design decisions, to allow for earlier, and broader, resilience consideration. Specific requests from DMCs are an important source of MDB-supported projects. By the time these requests are made, however, important project parameters have often already been set. In many cases, private sector projects come to MDBs at a particularly advanced stage, further narrowing the scope of possible options for assessing and improving resilience. Taking a more proactive approach to scoping projects right from the start unlocks additional opportunities for MDBs to optimize for resilience.

MDBs can achieve this shift by integrating asset-specific projects with broader strategic planning initiatives. MDBs support their DMCs through asset-specific planning and finance, as well as the development of sector or national strategies. However, a literature review of MDB activities indicated that project-specific initiatives are not systematically linked with strategic planning support.[3] Therefore, opportunities to promote strategic resilience priorities through individual projects supported by MDBs can be missed. Initiatives like the City Resilience Program of the World Bank (Box 1), AIIB's Sustainable Cities Strategy, and the EBRD Green Cities program, as well as country partnerships like those of ADB with Fiji, are existing examples of such integration, which could be scaled up further to achieve more consistent and systematic alignment between strategic and asset-specific support from MDBs.

Box 1: The World Bank's City Resilience Program

The program is aimed at empowering cities to make investments that promote resilience to disaster risks and to access financing for this purpose. It proposes a shift from sector priorities toward an integrated, spatially informed assessment of needs and priorities. To facilitate the shift, it makes geospatial tools available, shares case studies, and develops knowledge product workshops to apply lessons learned to the local context.

A tool called City Scan provides relevant spatial and socioeconomic information to enable city-level decision-makers to identify needs and opportunities for infrastructure investments that support the broader strategic development objectives of the city. As part of upstream support for mobilizing financing, the program also offers legal and capacity analysis, financial and regulatory analysis, and transaction advisory services.

Source: World Bank. 2020. *City Resilience Program*. Brief. 21 December.

In addition, MDBs can help to establish a clear picture of the disaster risk context in a country, which can be consistently used in planning various projects. The output from the multihazard risk assessment done in Tonga by ADB (Box 2), for instance, gives stakeholders a broad understanding of disaster risks across the country, which can be considered in early stages of site selection and planning, even before project-specific risk assessments are made. The World Bank and ADB published climate risk profiles for countries in South Asia, East Asia, and the Pacific. These reports, aimed at a broad range of development practitioners, provide an analysis of countries' climate characteristics, including projections and high-level assessments of natural hazard vulnerability (ADB n.d.[g]).

Box 2: ADB's Multihazard Climate and Disaster Risk Assessment for Tongatapu

Tonga, one of the most disaster-prone countries in the world, wished to conduct a multihazard risk assessment for one of its islands, with the help of ADB. Using national and regional datasets, ADB identified and mapped out key assets across the island of Tongatapu against current and future climate risks. The intent was to identify opportunities to effectively improve resilience in at-risk areas, and direct investments toward safer areas. This assessment was complementary to some other support measures extended to Tonga.

The project provides an example of how holistic risk assessments can be used by MDBs—and other stakeholders—to proactively identify resilience-building project opportunities before specific project locations or designs have been fixed. In many cases, this is the most efficient window for implementing resilience measures.

ADB = Asian Development Bank, MDB = multilateral development bank.
Source: ADB. Regional: Pacific Disaster Resilience Program (accessed January 2022); and Tonga and ADB (accessed January 2022).

---

[3] ADB, n.d.[e]; IDB, n.d.[h]; AIIB, n.d.[d]; EBRD, n.d.[h]; EIB, n.d.[e]; ISDB, n.d.; AfDB, n.d.[h]; World Bank, n.d.[j]

Opportunity 2: Improve integration of system-wide considerations within project-level risk assessments

Challenge

Risk assessments often focus on asset-level resilience considerations, and fail to capture system-wide interdependencies and cascading impact. Infrastructure systems are interdependent, and natural hazards can cause cascading risks of disruption. These cascading effects, which can ripple across entire infrastructure systems and economies, can far exceed the value of the asset itself. For instance, a study in Viet Nam concluded that disruption at one of the country's leading railways could lead to a cost of $2.6 million a day, because of significant cascading effects associated with the disruption in freight exchanges across regions. The disruption implies losses on both the demand side, as final goods cannot reach the relevant markets, and the supply side, because producers cannot access commodities for production. Rerouting costs and indirect effects on other firms and industries, as subsequent trade relationships are disturbed, also contribute to the total economic loss (Oh et al. 2019, cited in ADB 2021e). These considerations are not captured in risk assessments, which focus on individual infrastructure assets in isolation. This deficiency can be due to the complexity involved in modeling entire infrastructure systems, especially for large geographic areas, combining several layers of governance and data sources.

MDB expertise and resources are needed to foster resilience through system-wide planning; however, this knowledge is not yet systematically leveraged in project-specific planning. Through their involvement in sector and national development planning, MDBs can offer important expertise in system-wide resilience considerations, depending on the stakeholders interviewed. However, according to the literature review and several experts interviewed, MDB infrastructure finance is typically provided for individual infrastructure assets. Risk assessments associated with such finance therefore also focus on the asset itself. This asset-level approach can miss important interdependencies between the asset and the entire infrastructure system, and overlook the risk of cascading impact across the system and its users.

Multilateral Development Bank Opportunity

Deploying MDB expertise to broaden the risk assessment scope from the asset-specific to the system-wide level can result in the leveraging of more effective opportunities to improve user resilience. To effectively identify these opportunities, project-level risk assessments must be systematically linked with system-wide views of risk. Such system-wide views of risk may already exist within MDBs, through the broader strategic support they frequently provide at the sector and national levels. Ensuring a systematic, consistent consideration of a system-wide view within asset-specific planning and risk assessment activities offers significant scope for scaling up further the impact of existing MDB expertise in fostering the development of resilient infrastructure systems.

An example of MDB involvement in risk assessments that accounts for interdependencies within a network is provided in Box 3.

**Box 3: The Inter-American Development Bank's Blue Spot Analysis
for the Road Network in the Dominican Republic**

The IDB has allocated $650,000 to support the Government of the Dominican Republic in developing a risk management system for its roads and implementing the HydroBID-Flood risk model to model threats posed by different water perils. This funding is also intended to contribute to the updating of the national road inventory and the quality diagnosis of bridges and corridors, through technical cooperation. The project includes a Blue Spot analysis of the road network, involving a systematic analysis of critical points across the network based on risk assessments and an evaluation of alternative adaptation measures, taking uncertainty into account. The IDB also aims to support the prioritization of interventions, as well as engagement with stakeholders to ensure future capability to incorporate resilience.

Assessing entire road networks is an example of the system-wide risk assessment approach, which allows for interdependencies to be considered in the assessment and subsequent project design.

IDB = Inter-American Development Bank.
Source: IDB. 2020. *Sustainability Report 2019*. Washington, DC.

Opportunity 3: Overcome data challenges through coordinated and consistent sharing of risk information

Challenge

Risk quantification depends on reliable data and technical capacity, access to which is limited in some developing countries. Risks are complex and evolving; models may therefore not capture uncertainties consistently and robustly if data quality is insufficient. Stakeholders pointed out that system-wide risks beyond damage to individual assets, and future evolutions of risk across the long life cycle of infrastructure investments, are particularly challenging to analyze in data-sparse environments. They observed that developing countries often face challenges around limited data availability, given the less advanced data collection and observation facilities in those countries. Moreover, broader technical capacity may be insufficient for a robust assessment of disaster risks within their infrastructure planning processes.

Multilateral Development Bank Opportunity

If shared effectively, data generated by MDBs through project-specific and strategic planning support can provide significant insights beyond their original application. MDBs regularly conduct projects specifically dedicated to understanding and analyzing risks within their DMCs, at different scales. There is therefore room for MDBs to pool these relevant data and modeling output to facilitate risk analysis across the MDBs themselves and other stakeholders—and avoid duplicating efforts.

MDBs can contribute consistently to centralizing available data by sharing their risk assessment resources on global data platforms, which, in turn, can facilitate global benchmarking. Publicly available and user-friendly global platforms offering data for risk assessments are an essential tool for identifying and dealing with data-sparse environments. MDBs can focus on providing more decision-relevant risk information to fill existing gaps. They can contribute to developing a centralized repository of risk assessments and resilience practices to retain and scale up their value beyond project-specific applications and stakeholders. Ensuring that these model outputs are clear, easy to interpret, and easy to connect to decision criteria is critical to securing successful uptake (An example of a global risk data platform can be found in Box 4). Data shared could include vulnerability data on key assets, exposure data, or hazard layers. Metrics estimating the impact of specific natural hazards, for example, by combining information about infrastructure assets with maps of population density, economic activity, and hazard information, would improve risk management planning.

Systematic data sharing can be reinforced via institutional arrangements between MDBs, as well as dedicated technical capacity support. Institutional coordination is needed to ensure that data are effectively shared. Dedicated expert staff within each organization can do the technical work required to collect relevant analysis, align format, and ensure accessibility. A similar approach has been undertaken within the joint MDB working group on climate finance tracking (more details can be found in Box 8).

Box 4: Global Risk Data Platform

The Global Risk Data Platform (formerly known as PREVIEW) is a user-friendly web-based GIS portal. It allows users to visualize, download, or extract data on past hazardous events, human and economic hazard exposure, and risk from natural hazards. It is composed of 60 datasets, covering 9 types of hazards, and can be used for risk assessments. A centralized repository can help in identifying critical gaps and prioritizing future investments in risk information.

So far, the Global Risk Data Platform has used only economic risk data from the World Bank, among the MDBs. Data output from other MDB projects could be integrated into such a platform. Considerable insight could thus be delivered, costs reduced, and duplicative efforts avoided, both for other projects and for broader planning and strategy development. These benefits could outweigh the marginal cost of making the data accessible.

MDB = multilateral development bank.
Sources: United Nations Environment Programme and United Nations International Strategy for Disaster Reduction. PREVIEW Global Risk Data Platform (accessed January 2022); G. Giulani and P. Peduzzi. 2011. The PREVIEW Global Risk Data Platform: A Geoportal to Serve and Share Global Data on Risk to Natural Hazards. *Natural Hazards and Earth System Sciences*. 11. pp. 53–66.

Opportunity 4: Track success and lessons learned through consistent post-project monitoring

Challenge

MDBs' overview of project challenges and successes is limited once the project falls outside their direct mandate. While MDBs use a defined disaster risk screening process during their initial project review, only limited approaches exist to determine the resilience of projects once they have been fully implemented. MDBs have adaptation monitoring schemes for their ongoing portfolio projects, but systematic monitoring stops once the projects close. A review of MDB practices indicated the existence of several institutionalized mechanisms—such as the IDB's portfolio-level sustainability reports in the transport sector (IADB 2018b)—for evaluating sustainability at the portfolio level while projects remain in the project pipeline. These mechanisms no longer operate once projects exit the pipeline. Stakeholders interviewed also indicated that project managers are encouraged to include maintenance plans in their designs, and that operations are monitored while the project is ongoing. However, after project closure, maintenance considerations tend to be out of mandate and scope for the financing MDB, and there is no systematic monitoring in place.

Multilateral Development Bank Opportunity

Defining clear processes to track progress will help MDBs in collecting best practices and gathering key lessons learned for future risk-informed planning. A more consistent approach to monitoring risk and resilience throughout implementation and assessing resilience achievements and gaps at regular intervals after project completion can help to ensure that resilience standards are consistently upheld and that opportunities for improvement are identified and considered in future project planning.

MDBs can share post-project evaluation results and lessons learned with their DMCs. The opportunities and lessons identified through consistent monitoring are relevant not only to MDB project implementation but to the wider DMC context as well. MDBs can ensure that these findings are systematically shared in country partnership strategies, or directly with relevant policy-makers and sector stakeholders, to support risk-informed governance at every level. MDBs can also share lessons learned on key topics of interest among themselves through a formalized procedure (see Opportunity 11 for more details).

3 Financing Mechanisms

This section reviews financial mechanisms used by MDBs to improve infrastructure resilience. These mechanisms can be grouped into two categories:

- **Resilient infrastructure finance**. These are investments in infrastructure assets before a disaster occurs. MDBs may either provide either direct financing for infrastructure, or support market development to crowd in private sector financing.

- **Disaster risk financing**. Disaster risk financing (DRF) is financial support for swift recovery and reconstruction after a disaster has occurred. DRF may be arranged either before a disaster occurs ("ex-ante" mechanisms) or after a disaster has struck ("ex-post" mechanisms). Similar to the MDB support for infrastructure finance, DRF may be provided either through direct funding or through the development of private sector risk transfer offerings.

Resilient infrastructure and disaster risk financing offerings, challenges, and opportunities are discussed in the following sections. An overview is provided in Figure 3.

Figure 3: Financing Mechanisms—Challenges and Opportunities

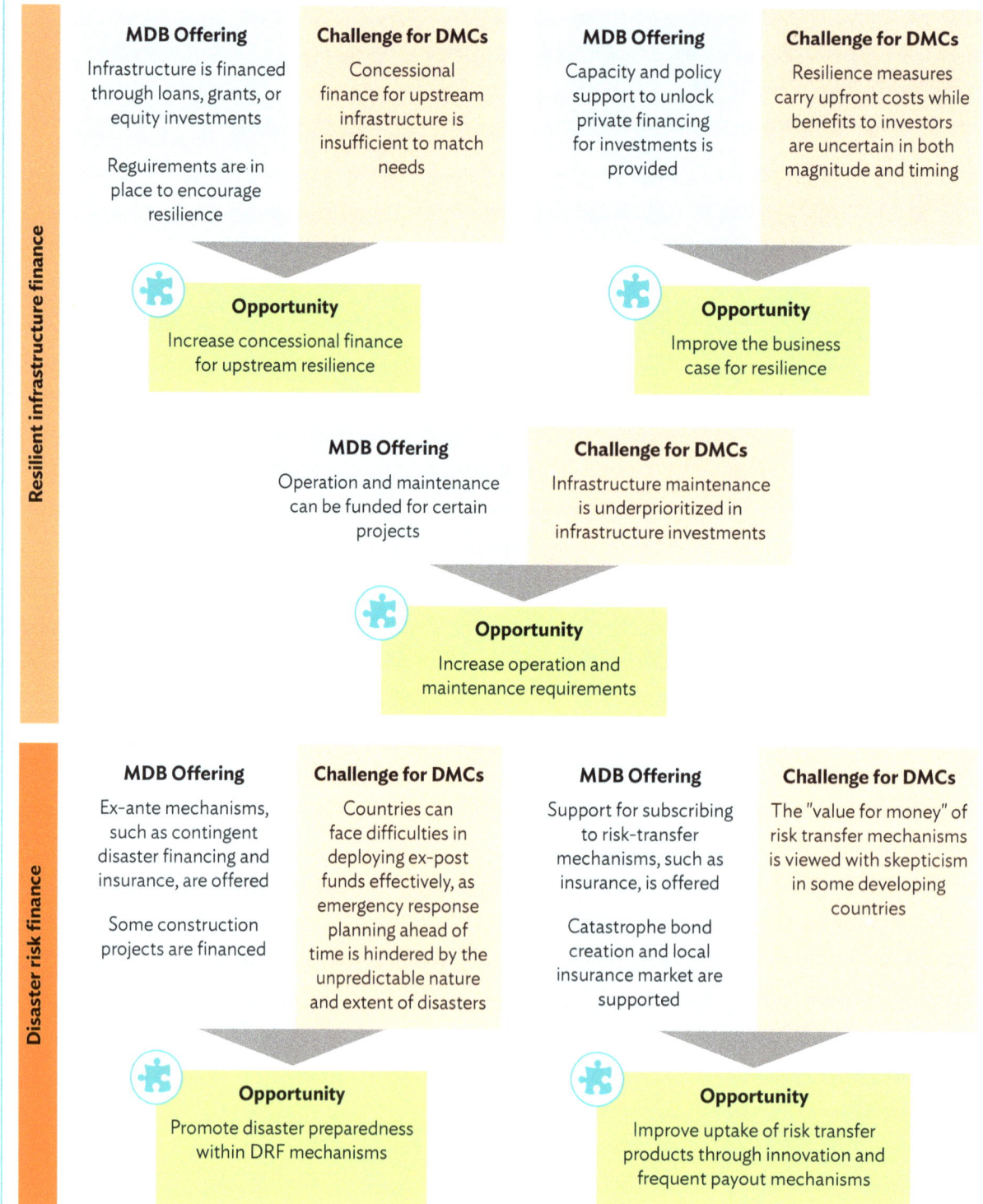

Resilient infrastructure finance

MDB Offering

Infrastructure is financed through loans, grants, or equity investments

Requirements are in place to encourage resilience

Challenge for DMCs

Concessional finance for upstream infrastructure is insufficient to match needs

Opportunity

Increase concessional finance for upstream resilience

MDB Offering

Capacity and policy support to unlock private financing for investments is provided

Challenge for DMCs

Resilience measures carry upfront costs while benefits to investors are uncertain in both magnitude and timing

Opportunity

Improve the business case for resilience

MDB Offering

Operation and maintenance can be funded for certain projects

Challenge for DMCs

Infrastructure maintenance is underprioritized in infrastructure investments

Opportunity

Increase operation and maintenance requirements

Disaster risk finance

MDB Offering

Ex-ante mechanisms, such as contingent disaster financing and insurance, are offered

Some construction projects are financed

Challenge for DMCs

Countries can face difficulties in deploying ex-post funds effectively, as emergency response planning ahead of time is hindered by the unpredictable nature and extent of disasters

Opportunity

Promote disaster preparedness within DRF mechanisms

MDB Offering

Support for subscribing to risk-transfer mechanisms, such as insurance, is offered

Catastrophe bond creation and local insurance market are supported

Challenge for DMCs

The "value for money" of risk transfer mechanisms is viewed with skepticism in some developing countries

Opportunity

Improve uptake of risk transfer products through innovation and frequent payout mechanisms

DMC = developing member country; DRF = disaster risk financing; MDB = multilateral development bank.
Source: Vivid Economics.

Current Multilateral Development Bank Offerings

At the project level, MDBs support investments in resilient infrastructure through traditional instruments tied to resilience requirements. Infrastructure finance enables infrastructure investments and upgrades, and represent a key opportunity for MDBs to promote infrastructure resilience. MDBs usually embed resilience requirements in traditional infrastructure finance instruments, such as concessional loans, grants, and equity investments.

MDBs offer dedicated sources of funding specifically for investments in resilience. They also make available mechanisms dedicated to financing resilience projects in particular. These can take the form of climate resilience funds. Green bonds, especially green resilience bonds, are another form of financing used specifically for investments in infrastructure resilience (Box 5). Such dedicated funding may, however, be primarily or exclusively targeted at improving resilience to climate hazards, leading to a reduced focus on non-climatic hazards such as earthquakes.

MDBs also provide capacity and policy support to unlock private financing for investments. MDB capacity support for capital market development involves financial reform, domestic bond market development, and the creation of an enabling environment that incentivizes the private sector to invest in infrastructure resilience. They likewise support the structuring of green bonds or other financial instruments to finance infrastructure investments.

MDBs offer both ex-ante and ex-post disaster risk finance support to increase liquidity after emergencies and foster resilient long-term reconstruction.

- Ex-ante mechanisms include direct financial support for several insurance schemes and pools, such as the InsuResilience Global Partnership and the Caribbean Catastrophe Risk Insurance Facility. This support can cover temporary or partial premium payments for lower-income DMCs, which may not purchase insurance coverage otherwise. In addition, technical support in structuring and accessing insurance is also provided, and contingent credit solutions are available as well to member states, usually accompanied with governance criteria.
- Ex-post response is supported by MDBs through a combination of technical assistance and funding. MDBs contribute to post-disaster needs assessments via technical assistance programs. They also provide immediate liquidity after a disaster, in the form of emergency loans or crisis response windows, which provide concessional finance or grants within days of a natural hazard event. Some MDBs offer options for restructuring ongoing project financing following disruptions due to disasters.

MDBs also aim to finance projects that support risk-informed development or the process of "building back better" after a disaster. Building back better (BBB) describes the repair or reconstruction of individual assets or entire systems in a way that improves their resilience beyond pre-disaster levels. This implies that response efforts surpass "back to normal" functioning and incorporate lessons learned from previous damage to enhance resilience, effectively using post-disaster finance to improve pre-disaster preparedness for future disasters.

Opportunities for Future Multilateral Development Bank Offerings

Opportunity 5: Increase concessional finance for upstream resilience

Challenge

Concessional finance for upstream infrastructure resilience is insufficient to match adaptation needs, making it challenging for developing countries to fund the required improvements. Climate finance, which includes transfers and concessional loans, designates finance dedicated to funding climate change mitigation measures and decreasing vulnerability to climate risk. Annual flows of $100 billion in climate finance for developing countries were pledged during the 2021 United Nations Climate Change Conference (COP26); half of this amount was to be spent on adaptation efforts (United Nations Framework Convention on Climate Change [UNFCCC], n.d.; UNFCCC COP26 2021a). In 2020, the MDBs collectively provided $16.1 billion in adaptation finance to their DMCs—24% of their total climate finance—and committed to increasing this amount to $18 billion by 2025 (AfDB et al. 2021b). However, adaptation costs are currently expected to range from $140 billion to $300 billion per year by 2030, with a further increase foreseen by 2050 (Antonich 2020). Infrastructure investment worldwide could face a funding gap of $15–$20 trillion by 2040, according to current trends, as public and private funds fall short of the total infrastructure needs of $94 trillion (Panwar, n.d.; Infrastructure Outlook, n.d.).

Moreover, resilience is considered a technical due diligence requirement for investments rather than a core goal, thereby restricting the scope of resilience projects considered for concessional finance. Infrastructure assets provide essential services to communities and economies, and are therefore designed with the goal of optimizing these services during normal, day-to-day operations. Additional design requirements to improve disaster resilience may be introduced through mandated risk screening processes, but are typically not central to the original project design. This implies that resilience projects may not be considered for concessional climate finance unless they can credibly demonstrate their ability to engender additional mitigation or adaptation benefits.

Multilateral Development Bank Opportunity

MDBs can increase the volume of financing dedicated to infrastructure investments for disaster resilience. Disaster-resilient infrastructure has an important role in fostering broader economic and societal resilience. MDBs can provide more concessional finance for infrastructure projects that recognize this role and are explicitly designed to promote disaster resilience as their central purpose. Offering concessional finance for such projects can help to shift the perception and treatment of resilience as an additional cost, and make it a core motivation for project selection and design.

MDB concessional finance should be channeled primarily into projects that enhance user resilience, to avoid distorting private sector incentives and maximize net social benefits. When designing concessional finance instruments for infrastructure resilience, MDBs should differentiate between asset resilience measures and improvements made to increase user resilience. Finance channeled into the former could distort investment incentives so that infrastructure owners are induced to prioritize asset- over system-level considerations. On the other hand, if funding is targeted instead at projects enhancing user resilience, particularly those that would not take shape without additional concessional finance streams, net social benefits can be maximized.

Associated measures that differentiate between infrastructure resilience and user resilience investments can be designed to track disaster resilience projects. The tracking could build on existing schemes like the annual joint MDB report on climate mitigation and adaptation finance (AfDB et al. 2021b). Tracking resilience investments and their outcomes can contribute to building evidence of the benefits unlocked by resilience and, in turn, strengthen the business case for such investments.

Opportunity 6: Improve the business case for resilience

Challenge

Unlocking private sources for finance is critical to the disaster resilience of infrastructure systems, given the insufficiency of public funding to fill the infrastructure resilience investment gap (Panwar, n.d.), yet the business case for private investments is perceived as weak. As disaster risks are set to increase further over time, even larger financing will be needed to close the widening investment gap. However, resilience is sometimes seen as an undesirable extra cost by investors. The upfront costs can be large and some investments have limited recovery potential through user charges. The benefits, on the other hand, are generally analyzed and presented incompletely in the private sector, stakeholders pointed out. Their timing is uncertain and they do not always translate into economic transactions, as they are difficult to monetize and often support public goods (Tall et al. 2021), making the projects less than attractive to investors under pressure to deliver short-term returns. Interviewees also noted the differentiated levels of resilience across investments, depending on the type of financier. While individual projects financed by MDBs tend to achieve good levels of resilience to disasters, they said, system-wide resilience can remain limited by less resilient assets within the system that are funded from other sources.

Some mechanisms for unlocking private investment, such as green bonds, are difficult to set up and not always dedicated to resilience investments. Several developing countries face high barriers in setting up green bonds, and despite substantial demand, green bond issuance in developing countries has remained low, making up only 0.5% of total bond issuance (Deschryver and de Mariz 2020). Moreover, according to the Climate Bonds Initiative, only 3%–5% of global green bond proceeds in 2018 were channeled toward climate adaptation and resilience (CBI 2018).

Multilateral Development Bank Opportunity

MDBs can support private investment by improving the business case for resilience and expanding the scope of project appraisal. MDBs can raise awareness of disaster risks and encourage their integration into commercial lending, due diligence, and risk analysis. The cost of not investing in resilience would thus become clearer. MDBs can provide standardized frameworks and technical assistance for the evaluation of a holistic range of benefits. The evaluation could be based, for example, on the "triple dividend" of resilience (Tanner et al. 2015)—reduced damage during a disaster, unlocked economic activity, and developmental co-benefits. Models incorporating these benefits are only starting to emerge and are not yet widely shared.

MDBs can support their DMCs in designing and implementing revenue models that incorporate a wide range of resilience benefits and reward investors for implementing resilient design. Standardized revenue models, which consider resilience costs and benefits, can help create new incentives for private sector commitment. These mechanisms can ensure that the party bearing the cost of investing in resilience also has access to the associated returns. In turn, this can encourage private investors to take resilience into consideration, even in projects completely outside the MDB project pipelines. MDBs can support the design of these models and play an advisory role in their respective DMCs to ensure implementation and encourage uptake.

Some initiatives to reward resilience investments are already in place and can be shared more widely and expanded on. The AfDB's Adaptation Benefit Mechanism (ABM), which captures and certifies the social, economic, and environmental benefits of adaptation investments, is designed to help investors understand the full benefits of resilience but is still in its pilot phase (AfDB, n.d.[a]). MDBs could replicate and expand on such initiatives, to give investors a clear picture of the concrete implications of their investments in resilient infrastructure. Schemes like the ABM can have positive reputational consequences for investors that invest in the most valued projects. They could also go hand in hand with other incentive mechanisms that accurately reflect the cost–benefit trade-off, such as subsidies or tax breaks, with positive impact on the business case for resilience. Alternatively, the Coalition for Climate Resilient Investment (CCRI), an initiative funded in part by five MDBs, created the Physical Climate Risk Assessment Methodology (PCRAM), which aims to put a price on climate risk and reflect the financial benefits of resilience in asset valuations. This methodology analyzes the impact of climate change on portfolios, suggests resilience options, and details associated costs and benefits (CCRI, n.d.). The results from the analysis can then be included in investment appraisals. MDBs can build on such initiatives, to provide investors with a vision of the concrete impact and costs of disaster risks on their investments.

MDBs can also unlock further private investment by supporting and creating dedicated financing mechanisms, such as green resilience bonds. MDBs can contribute to developing expertise within their DMCs and promote the development of national green bond schemes aimed at improving resilience, as well as issue their own green bonds with resilience criteria. These bonds can increase the availability of funds specifically for resilience projects, making it easier for project managers to deal with the upfront cost of investing in resilience. They can also encourage the inclusion of resilience components in the design of projects for which disaster risk was originally not considered, to benefit from the funds. An example of a green bond with resilience criteria, put in place by the EBRD, is presented in Box 5.

Box 5: The European Bank for Reconstruction and Development's Climate Resilience Bonds

Private financing of resilient projects needs to increase to match adaptation needs. To address this issue, the EBRD launched its Green Bonds in 2019. Among these was a 5-year, $700 million Climate Resilience Bond, intended to finance projects focused on climate-resilient business and commercial operations or climate-resilient agriculture and ecological systems. EBRD's Environment and Sustainability, Banking, Treasury, and Legal departments set the eligibility criteria for projects. To be financed through this means, projects must fit in with the Climate Resilience Principles, which include both asset-level and system-level resilience considerations.

As of May 2021, the Climate Resilience Portfolio contained €1.4 billion in operating assets, €1.25 billion of which was dedicated to climate-resilient infrastructure. The majority of the operating assets in the overall portfolio have been set aside for investments in the transport sector, followed by energy and agribusiness.

Green resilience bonds can serve as a mechanism for drawing in private finance for infrastructure resilience projects. A clear set of criteria allows for projects with high adaptation benefits to be prioritized for access to finance.

EBRD = European Bank for Reconstruction and Development.
Source: Climate Bonds Initiative. Climate Resilience Principles (accessed January 2022).

Opportunity 7: Increase operation and maintenance requirements

Challenge

Maintenance is underprioritized in investments. This means that the ability of infrastructure to withstand disasters may diminish over time even if resilience measures are implemented up front. Maintenance and investment upgrades have a vital role in ensuring continuous resilience and avoiding expensive repairs as well as capital-intensive construction needs in the long run (Rozenberg and Fay 2019; Kornejew, Rentschler, and Hallegatte 2019). However, constrained short-term budgets, regulatory asset base mechanisms, or contractual arrangements separating the infrastructure owner and operators often lead to maintenance spending being underprioritized.

The upkeep of infrastructure assets is insufficiently enforced and monitored, as investment in maintenance tends to fall outside the scope of MDB mandates. Large-scale project finance is at the core of MDB mandates; smaller-scale maintenance opportunities are less prevalent in current MDB activities. While the development of asset maintenance plans is often part of MDB project requirements, their enforcement is typically not monitored once the project is completed.

Multilateral Development Bank Opportunity

MDBs can enforce stronger maintenance requirements on new and rehabilitated assets, through enhanced monitoring and dedicated funding options. Maintenance options offer generally favorable returns in promoting resilience. Capital costs can increase by 50%–60% (Rozenberg and Fay 2019) because of inadequate maintenance; conversely, $1.50 of investment costs can be saved for every additional $1 spent on road maintenance (Kornejew, Rentschler, and Hallegatte 2019). To ensure that capital resilience improvements remain effective at withstanding disasters over time, MDBs can take the following actions to encourage and include maintenance in project management:

- Oversee not only the development of asset maintenance plans but also their execution across the entire asset life cycle.

- Expand their offering by including maintenance-specific funding routes, such as requirements to scan for maintenance or upgrade options before financing for a new asset is granted.

- Explicitly consider maintenance cost in early project design, and highlight the benefits of more resilient design options in reducing long-term costs across the asset lifetime.

Resilience can also be supported if MDBs encourage the adoption of "best practice" operational mechanisms. Operational models that effectively prioritize user needs can allow network operators to respond to changing circumstances, thereby enhancing operational resilience. If operations are designed with built-in flexibility, they are more likely to continue to provide an acceptable level of infrastructure services after a disaster occurs. Examples of effective mechanisms for enhancing operational resilience are the increased use of information technology systems, regular monitoring, and investment in local-level training. MDBs can play a crucial role in establishing such operational measures at project inception, and encouraging continuous practice throughout the asset life cycle.

Opportunity 8: Improve uptake of risk transfer products through product innovation and frequent payout mechanisms

Challenge

The "value for money" of risk transfer mechanisms is viewed with skepticism in some developing countries. As decision-makers and infrastructure operators must work within tight budget constraints, the opportunity cost of risk transfer premiums can be considered too high. This is particularly true for disaster-prone countries, which face higher premiums proportional to the increased likelihood of the insurance policy being triggered. Stakeholders interviewed also indicated that the prospect of insurance payouts is met with mistrust among some policy-makers. A better understanding of the value of risk transfer and more tailored solutions to existing concerns in developing countries are required to increase the uptake of ex-ante risk transfer and reduce reliance on ex-post emergency funding.

Multilateral Development Bank Opportunity

MDBs can leverage and drive innovations in product design to increase the reach and relevance of insurance solutions to specific challenges faced in developing countries. New and improving technologies, such as satellite imagery, enable insurance mechanisms, such as parametric triggers, to become increasingly robust and scalable. In addition, emerging innovations allow for the coverage of a broader set of hazards and losses, such as the economic impact of coral reef degradation (Mesoamerican Reef Fund, InsuResilience Solutions Fund, and Willis Towers Watson 2021). By mapping available solutions to existing challenges in developing countries, and designing products tailored to the responsibilities and priorities of individual stakeholders, MDBs can promote the relevance of risk transfer to the different roles of infrastructure owners and operators, from the national to the local level.

Mechanisms that encourage frequent payouts, even if limited in magnitude, can help to promote the value proposition of risk transfer mechanisms. Risk transfer designs that combine large-scale payouts for severe events with small payouts for more frequent events can present an opportunity to foster understanding and increase political acceptance. Successful small payouts have been delivered by the African Risk Capacity Insurance Company Limited (ARC), on the basis of premium support offered by AfDB, as detailed in Box 6. According to stakeholder interviews, these payouts have contributed greatly to strengthening the value proposition of insurance schemes in the eyes of policy-makers.

Box 6: African Risk Capacity Insurance Company Limited

The African Risk Capacity Insurance Company Limited (ARC) is a specialized agency of the African Union that provides parametric insurance solutions to its members. Risk is pooled across the region to diversify the risk portfolio and reduce associated risk transfer costs. ARC is supported by the AfDB via the Africa Disaster Risks Financing (ADRiFi) program, which enables countries to access risk transfer services from ARC through premium support.

With ADRiFi's premium support, Madagascar had access to drought insurance in 2019 and/or 2020 and in 2020, receiving a total payout of $2.13 million from the ARC. Even though the payout remained far below the entire economic impact of the drought, it significantly improved the perceived value of insurance in the country and the region, and contributed to the support of 600,000 vulnerable people.

Structuring insurance mechanisms to trigger small, frequent payouts can be an effective way of demonstrating the value of insurance, by delivering examples of successful payouts even in the absence of the most severe, but infrequent, disaster events. Moreover, when directed toward the right assets, small-magnitude payouts can have a large impact on the affected population and contribute effectively to market development.

AfDB = African Development Bank
Source: ARC. 2021. Settling a Claim in Madagascar Proves the Value of Insurance. 13 December.

In addition, MDBs can increase risk transfer penetration by mandating insurance coverage for infrastructure assets financed by them. The purchase of disaster risk insurance for assets supported by MDBs is currently not consistently mandated across all MDBs and project types. Introducing such insurance requirements can offer an additional mechanism for improving understanding of the value of risk transfer and ensuring that premium costs are integrated into the early stages of project planning.

Opportunity 9: Promote disaster recovery planning within DRF mechanisms

Challenge

Developing countries can face difficulties in deploying ex-post funds effectively, as emergency response planning ahead of time is hindered by the unpredictable nature and extent of disasters. Stakeholders interviewed for this report indicated that even if funds are disbursed rapidly after a natural hazard, developing countries can lack the capacity to rapidly deploy funding to effectively support response and relief efforts. Interviewees highlighted a particular need for post-disaster capacity building, through such means as the creation of appropriate disaster risk management committees. The uncertain magnitude, source, and timing of funds hinders efficient planning ahead of a disaster, and the limited capacity of individual operators to respond can result in overwhelming dependence on intervention from national governments.

Multilateral Development Bank Opportunity

MDBs can tie DRF mechanisms to conditions that promote improved response capacity in the aftermath of a disaster. Both ex-ante and ex-post DRF mechanisms can be linked to conditions that help developing countries to access and use funds received in the aftermath of a disaster. These mechanisms can include requirements for emergency response planning, institutional frameworks, or specific policy reforms for members, effectively combining improved ex-ante resilience with increased ex-post response capacity. An example of a contingent credit mechanism tied to policy requirements and continuous monitoring can be found in Box 7.

In particular, MDB-led DRF mechanisms offer an opportunity to establish clear risk ownership and incentivize resilience-building investments before a disaster strikes. The determination of risk ownership is thus an essential first step in setting incentives for resilience. Regulatory and financial levers to incentivize risk management may be difficult and expensive to enforce, resulting in many developing countries struggling with low compliance. MDBs can help to address this issue by allocating clear responsibilities for risk ownership and recovery planning to beneficiaries of DRF mechanisms.

Box 7: Contingent Credit for Natural Hazards in Peru

The IDB allocated a $300-million contingent loan to Peru in 2013, to be disbursed rapidly in case of an earthquake.

The loan was made to improve the country's ability to respond to disasters, through immediate funding and upstream policy reforms. Access to the contingent credit facility is conditional on the development of certain risk management mechanisms, including the creation of the Comprehensive Natural Disaster Risk Management Program. A range of indicators of disaster risk management is also monitored yearly.

Policy requirements for contingent credit offer an opportunity to combine ex-ante resilience with increased ex-post response capacity, and foster improved long-term planning initiatives. They can also help improve countries' capacity to deploy disaster response funds (IDB 2013b; stakeholder input).

IDB = Inter-American Development Bank
Source: *IDB*. 2013. Peru Improves Ability to Respond to Natural Disasters with IDB Support. 13 December. Washington, DC.

4 Knowledge Building

This chapter describes the procedures through which MDBs currently provide support for knowledge creation and integration among their DMCs, as well as the related challenges and opportunities to overcome them. Informed disaster risk decision-making requires accurate and up-to-date knowledge regarding disaster risks. MDBs play a key role in disseminating information across their DMCs, by creating relevant knowledge products to support infrastructure disaster resilience and facilitating the integration of new knowledge in the DMCs through capacity building. MDBs are uniquely placed to support their members in overcoming the challenges they are facing, such as constantly evolving risks (Weichselgartner and Pigeon 2015; GFDRR 2014; Tran and Boyland 2022), as well as tracking progress. Figure 4 presents an overview of the challenges and opportunities analyzed in this section.

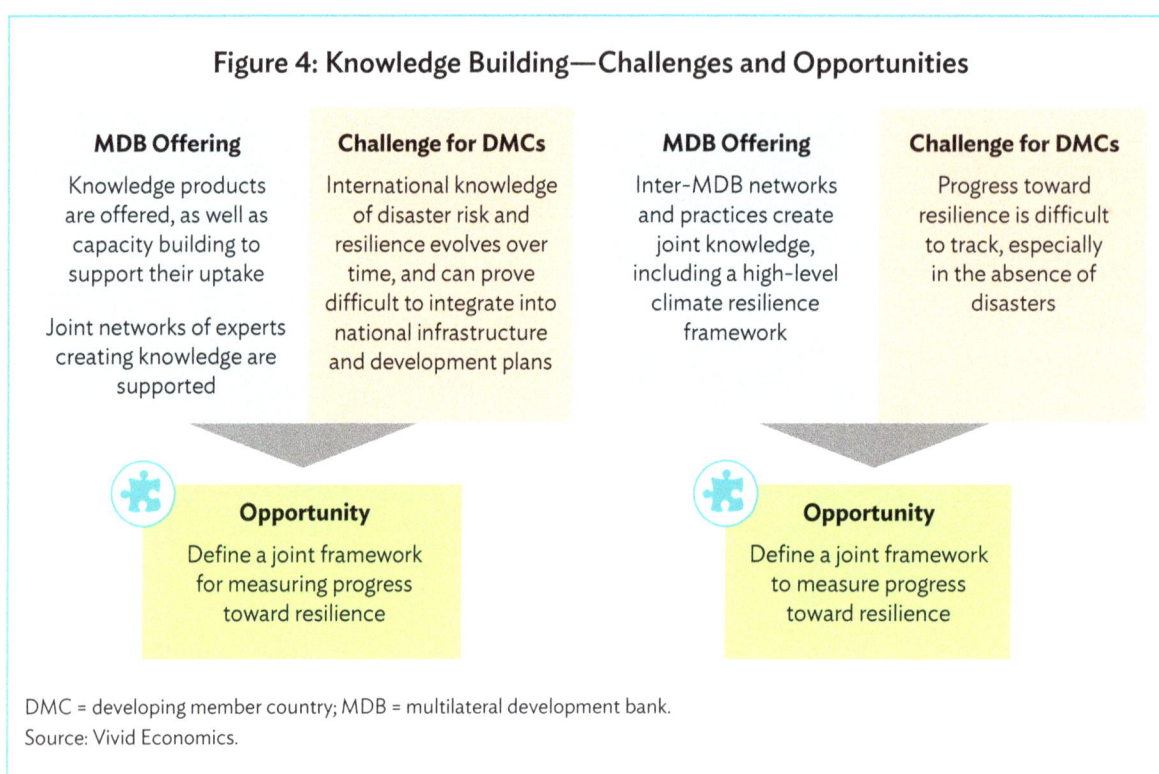

Figure 4: Knowledge Building—Challenges and Opportunities

MDB Offering	Challenge for DMCs	MDB Offering	Challenge for DMCs
Knowledge products are offered, as well as capacity building to support their uptake Joint networks of experts creating knowledge are supported	International knowledge of disaster risk and resilience evolves over time, and can prove difficult to integrate into national infrastructure and development plans	Inter-MDB networks and practices create joint knowledge, including a high-level climate resilience framework	Progress toward resilience is difficult to track, especially in the absence of disasters

Opportunity
Define a joint framework for measuring progress toward resilience

Opportunity
Define a joint framework to measure progress toward resilience

DMC = developing member country; MDB = multilateral development bank.
Source: Vivid Economics.

Current Multilateral Develpment Bank Offerings

MDBs currently offer several knowledge products and associated training for their uptake in DMCs. MDBs produce guides on infrastructure-related topics, including promoting private sector investment, making infrastructure projects more disaster-resilient, and improving understanding of climate change impact on specific workstreams. They also produce open methodologies and data for risk assessments, as well as early warning systems. In parallel, MDBs regularly conduct capacity-building initiatives to further support the dissemination of this knowledge.

MDBs host networks of practitioners to share experiences, challenges, and best practice, and participate in international forums on disaster risk. For example, initiatives like the Disaster Risk Management Network, a network of policy-makers from several risk management institutions, benefit from MDB support (IDB, n.d.[d]). MDBs also contribute to global conferences on disaster risk in general and climate hazards in particular, thus informing and shaping global knowledge and legislation. For instance, during COP26, the MDBs delivered a joint statement on their collective engagements and ambitions related to compliance with the Paris Alignment (UNFCCC COP26 2021b).

MDBs also foster knowledge through joint networks and practices. These networks promote knowledge sharing and alignment across MDBs on key topics and emerging challenges. For example, there are several such initiatives on topics related to climate change, including the annual Joint Report on Multilateral Development Banks' Climate Finance (Box 8).

Opportunities for Future Multilateral Development Bank Offerings

Opportunity 10: Increase integration of new knowledge through formal structured collaboration

Challenge

The evolving nature of knowledge of resilience and disaster risks creates challenges for continuous integration in project pipelines. International knowledge and consensus on resilience, disaster risk, and climate change is constantly evolving (Tran and Boyland 2022). For the DMCs to benefit from the most efficient solutions, MDBs need to keep pace with this new knowledge systematically. However, stakeholders interviewed indicated that there can be gaps and delays in adopting the necessary expertise. For instance, expertise in procuring and scaling up nature-based solutions (NBSs) for infrastructure projects could be increased, according to interviewees.

Multilateral Development Bank Opportunity

Joint MDB initiatives can ensure systematic collaboration and alignment, in particular on emerging topics of joint importance. A systematic framework of knowledge exchange can foster integration and alignment, as it creates a standard procedure for sharing findings. It can ensure that MDBs have an overview of current knowledge, and can easily access the most up-to-date expertise and integrate it into their practices. Complementarities between analyses and existing gaps in literature can also be highlighted this way. This is particularly relevant when it comes to sharing new and emerging topics of interest, such as NBSs or private sector engagement, across MDBs. Joint MDB climate finance initiatives have fostered progress on this important topic over the past years (Box 8) and could be used as a blueprint for initiating further collaboration in other areas, such as disaster risk, resilience, or NBSs.

Box 8: Joint Multilateral Development Bank Working Group on Climate Finance

In 2012, the first Joint MDB Report on Mitigation Finance, providing an overview of investments in climate change mitigation during the preceding year, was published by six MDBs (AfDB, ADB, EBRD, EIB, IDB, and the World Bank). The MDBs followed slightly different approaches to reporting and classifying climate finance, but created a joint working group to harmonize the methodologies in future publications.[a]

In 2015, joint methodologies for tracking mitigation and adaptation finance—Common Principles for Climate Mitigation Finance Tracking and Common Principles for Climate Change Adaptation Finance Tracking—were published by the working group in collaboration with the International Development Finance Club (IDFC).[b] The tracking principles for mitigation finance were updated in 2021, while the tracking principles for adaptation are now under review.[c]

As of 2020, nine MDBs—the eight reviewed in this report, as well as the New Development Bank (NDB)—annually track their mitigation and adaptation finance jointly and present the results in the Joint Report on Multilateral Development Banks' Climate Finance.[d] In addition, a group of representatives of the IDFC member banks also implements these jointly developed principles for tracking climate finance.

Seven MDBs in the working group (ADB, AfDB, AIIB, EBRD, EIB, IDB, and IsDB) likewise developed the Common Framework for Climate Resilience Metrics in Financing Operations in 2019. This document lays down a joint high-level framework for designing and implementing resilience measurements for MDB operations, and gives examples of metrics and their implementation, but does not set out a precise list of common tracking measure.[e]

The group is an example of a global network used to generate common standards and approaches. These initiatives offer the opportunity to further improve alignment on resilience activities and could be extended to other topics such as nature-based solutions, as well as drive progress on technical challenges, such as system-wide risk assessment.

ADB = Asian Development Bank, AfDB = African Development Bank, AIIB = Asian Infrastructure Investment Bank, EBRD = European Bank for Reconstruction and Development, EIB = European Investment Bank, IDB = Inter-American Development Bank, IsDB = Islamic Development Bank, MDB = multilateral development bank.

[a] AfDB et al. 2012. *Joint MDB Report on Mitigation Finance 2011*. Washington, DC: World Bank.
[b] AfDB et al. 2021. *2020 Joint Report on Multilateral Development Banks' Climate Finance*. London: EBRD.
[c] AfDB et al. 2021a. Common Principles for Climate Mitigation Finance Tracking. Version 3. 18 October. Frankfurt, Germany: IDFC; AfDB et al. 2021. *2020 Joint Report on Multilateral Development Banks' Climate Finance*. London: EBRD.
[d] AfDB et al. 2021. *2020 Joint Report on Multilateral Development Banks' Climate Finance*. London: EBRD; United Nations Framework Convention on Climate Change 26th Climate Change Conference. 2021. *Progress Report: Multilateral Development Banks Working Together for Paris Alignment*. 3 November.
[e] AfDB et al.2019. *A Framework and Principles for Climate Resilience Metrics in Financing Operations*. Washington, DC: IADB.

Clear structures for collaboration between disaster risk "knowledge" experts and operational teams within MDBs can accelerate the integration of knowledge into project operations. In addition to initiatives across different MDBs, continuous and consistent exchange between different teams within MDBs can help to define and effectively drive a shared vision across the organization. While collaborations are frequently established for project-specific purposes, they often rely on the initiative of individuals and are not continuously upheld beyond the scope of the project, according to stakeholders interviewed for this report. A more structured model of collaboration between knowledge experts and operational teams can help keep projects fully informed with the latest knowledge products, and enable rapid testing of those products in operational applications.

At the country level, MDBs can leverage long-term partnerships to promote institutional knowledge and continuous learning. The infrequent occurrence of disasters means that lessons learned at the sector and country level across developing countries are limited. Long-term partnerships with countries, regions, and sectors are essential in identifying key challenges and efficiently directing investments and delivering appropriate solutions. To maximize the benefits of investments made by MDBs in building these partnerships, it is critical that they build capacity that does not rely on individuals and is therefore preserved despite changes in stakeholders. It is also important that long-term initiatives do not operate in isolation but are systematically and proactively linked to project-specific activities to ensure that knowledge is shared and duplication of effort avoided.

NBSs are a specific example of an area of widespread interest in which increased collaboration could help to overcome capacity and knowledge gaps, which currently impede effective identification and scaling. NBSs are an area of joint interest among MDBs, as well as their DMCs. However, unlike hard engineering solutions, NBSs are still constrained by limited understanding of their design, procurement, and implementation, although these are active areas of ongoing research among many MDBs and their partners. Experts interviewed highlighted the lack of training for MDB project teams in the usability of these solutions. A joint initiative across MDBs could help to speed up the development of theoretical knowledge as well as practical guidance for project managers and stakeholders in members. In addition, limitations and difficulties associated with NBSs, such as the significant land area requirement, for which acquisition may be outside their mandate, could be examined further. Finally, existing solutions and expertise can be highlighted through the sharing of implementation examples, such as the project described in Box 9.

Box 9: Addressing Flood Risk through "Sponge Cities"

Flood frequency and related damage have increased in urban areas in the People's Republic of China (PRC) in recent years. As the population becomes more urban, this trend is likely to continue.

ADB allocated a $150 million loan to the PRC to improve flood and water management in Jiangxi province, and particularly in Pingxiang city. The project supports water and wastewater legislation, creates risk management partnerships, and aims to foster community empowerment and ecological river management supporting sustainable urban–rural sponge city development. The solutions developed to tackle flooding include flood plains protection, restoration of wetlands, and creation of wider green spaces along rivers. The wetlands neighboring rivers will be rehabilitated to cleanse the rainwater.

The results and expertise gathered from the Sponge City Project were presented by ADB at a conference on resilient infrastructure. Knowledge gathered from large-scale nature-based solutions (NBSs) projects could benefit from being systematically shared among MDBs. This type of initiative can contribute to developing expertise in NBSs operationalization within MDBs and among stakeholders, as well as incentivize standardized regulation.

ADB = Asian Development Bank; MDB = multilateral development bank.
Sources: ADB. 2015. *Jiangxi Pingxiang Integrated Rural–Urban Infrastructure Development: Final Report*. Consultant's report. Manila (TA 8451-PRC); G. Dwyer. 2017. Piloting "Sponge Cities" in the People's Republic of China. Project result/Case study. *ADB*. Manila. 13 January; S. Rau. 2021. Sponge Cities in the People's Republic of China—Evolution of ADB Support. Presentation at the Asian Development Bank Knowledge Events webinar Nature-based Solutions for Resilient Infrastructure. 15 July.

Opportunity 11: Define a joint framework for measuring progress toward resilience

Challenge

The infrequent occurrence of disasters makes it challenging to track and reward progress toward resilience. In the absence of disasters that explicitly test the ability of infrastructure systems to withstand and recover from catastrophic impact, it is difficult for members to track progress toward resilience. To effectively demonstrate improvements in disaster resilience before a disaster strikes, a standardized framework that can holistically measure resilience across its different dimensions (e.g., institutional capacity, risk assessment, emergency response planning) is required.

MDBs currently do not have a common framework for measuring disaster resilience, and this deficiency can hinder global tracking and scaling efforts. MDBs individually assess the resilience potential of their investments and can, on occasion, provide country-level evaluations of the disaster risk and resilience capacity of their member states. However, such assessments are not carried out systematically and consistently across MDBs. When conducted, they vary in both methodology and scope, as each MDB follows its own results framework. Similarly, risk assessment tools and standards vary across MDBs, making it difficult to compare risk and resilience across banks and regions.

Multilateral Development Bank Opportunity

A standardized resilience tracking framework across MDBs can help to mark progress and identify remaining gaps and priorities at the global level. A clear, measurable set of metrics for tracking resilience can direct action by MDBs and support policy-makers across the DMCs in continuously demonstrating the impact of investments in resilience. These metrics can provide MDBs with a common overview of resilience status, progress, and needs at the country or regional level, and thus allow them to evaluate progress and gaps globally. They will also enable MDBs to direct investments to the most relevant areas and projects, and prioritize those investments, by identifying regions where resilience investments are insufficient, as well as sectors where resilience efforts are concentrated. Likewise, an overview at the national level, provided systematically and in a user-friendly manner to policy-makers, will give them helpful support. Tracking measures can also help to structure and align the scope and output of risk assessments, and hence facilitate the collection of outputs on a unified platform (see Opportunity 3).

The proposed framework can incorporate indicators collected through other initiatives and can complement and expand on existing initiatives. Some elements of existing frameworks, such as IADB's Index of Governance and Public Policy (iGOPP) (Box 10), could be integrated into the proposed framework. The iGOPP aggregates national indicators, some of which are particularly pertinent to resilience evaluation. Among these relevant indicators (Lacambra et al. 2015b) is the existence of regulatory mandates to ensure that the public assets of sector and territorial entities are in place, or to carry out disaster risk reduction measures during the construction of public and private infrastructure projects. Indicators showing whether disaster risk is taken into account in national development planning are also relevant. The proposed framework would complement the Framework and Principles for Climate Resilience Metrics in Financing Operations (AfDB et al. 2019). While these existing principles provide MDBs with a common approach to evaluating the resilience impact of their investments (Box 8), the new set of metrics could continuously and consistently track resilience levels across different geographies and time periods, including intervals when no disaster events occur.

The common metrics for monitoring resilience across MDBs should give institutions sufficient scope to incorporate specific requirements and considerations based on their regional context. MDBs should aim to limit their common approach to high-level metrics, leaving the elaboration and application of indices to individual institutions. This reflects the wide diversity of institutional mandates, business models, strategic priorities, and operational contexts.

Box 10: Index of Governance and Public Policy in Disaster Risk Management

At the request of its members, a group of DRM experts from IDB and the academe developed an index for evaluating the presence of sufficient legal, institutional, and budgetary conditions for successful disaster risk management. Existing methodologies were taken into account in the creation of the Index of Governance and Public Policy (iGOPP).

This index allows the quantification of the extent to which a country's institutions are aligned with DRM objectives. It is based on both governance and risk management frameworks, and has been correlated with resilience performance as measured by human losses in case of disaster (Lacambra et al. 2015a, 2015b; Guerrero Compéan and Lacambra Ayuso 2020).

A consistent approach to defining and evaluating resilience through an index like the iGOPP offers opportunities to better target MDB initiatives and align them with a common goal. The introduction or adoption of such indexes could be an important step in advancing existing initiatives, such as the Framework and Principles for Climate Resilience Metrics in Financing Operations, jointly created by several MDBs.

DRM = disaster risk management; IDB = Inter-American Development Bank; MDB = multilateral development bank.
Sources: S. Lacambra et al. 2015. *Index of Governance and Public Policy in Disaster Risk Management (iGOPP): Application Protocol*. Washington, DC: IDB; S. Lacambra et al. 2015. *Index of Governance and Public Policy in Disaster Risk Management (iGOPP): Main Technical Document*. Washington, DC: IDB; R. Guerrero Compeán and S. Lacambra Ayuso. 2020. Disasters and Loss of Life: New Evidence on the Effect of Disaster Risk Management Governance in Latin America and the Caribbean. *IDB Working Paper Series*. No. IDB-WP-01126. Washington, DC: IDB.

5 Conclusion

MDBs already offer a large range of instruments to support their members in promoting disaster-resilient infrastructure systems. Specific offerings vary across MDBs, but generally include a range of capacity building, technical assistance, and financing options at the national, sector, and project levels. However, there are opportunities for MDBs to help decision-makers understand disaster risks better, prioritize investments, and build projects that foster system-wide resilience.

This report contributes toward a joint understanding of key opportunities for MDBs to further expand and scale up existing offerings. It examines current challenges faced by DMCs across Asia and the Pacific, and identifies options for MDBs to support their members in these specific areas. Opportunities are motivated by specific challenges faced by developing countries, and illustrated with examples of existing best practices that highlight possible routes to implementation.

Key opportunities and next steps for MDBs emerging from this report are centered on three key themes:

- integrating MDB support more consistently across the entire lifecycle and system of infrastructure assets, starting with early involvement in project conception and extending to systematic coordination with strategic national or sector planning;

- strengthening the business case for resilience among public and private stakeholders, through access to user-friendly risk information and targeted financing solutions, to elevate disaster resilience to core consideration in decision-making; and

- increasing collaboration within and across MDB teams to accelerate progress, alignment, and operational uptake in key areas of joint interest.

As disaster risks are increasing globally, these opportunities offer MDBs a pathway to supporting their members in meeting new resilience challenges. Without targeted action and as populations grow, climate change and unplanned urbanization will worsen existing resilience gaps and imperil the sustainability of past and future development across Asia and the Pacific and globally. MDBs can leverage the opportunities identified in this report to help address these challenges, and support their DMCs in building disaster-resilient infrastructure systems that unlock lasting economic growth and prosperity.

Appendix: Current Offerings of Individual Multilateral Development Banks

The following tables summarize each multilateral development bank's offering to support infrastructure disaster resilience in more detail. Each table combines findings from a review of the literature—sources are listed in the References section—and stakeholder input.

Table A.1: African Development Bank Offering

Risk-informed planning

- joint implementation of the ClimDev Special Fund, which is aimed at increasing capacity to generate weather data, climate change evidence, and implications
- promotion of early warning systems
- support for country-level risk profile development (ADRiFi), tied to some financing mechanisms (ARC)
- guidance in project design, including climate risk assessment (AfDB Integrated Safeguards System)
- requirement to address climate risks strategically in country strategy papers and regional strategy framework
- technical assistance, with emphasis on project preparation phase

Financing mechanisms

Resilient infrastructure finance	Disaster risk finance
Instruments • **loans and equity investments** **Market development** • **institutional support, knowledge dissemination, and specific projects carried out to mobilize private finance** through the Capital Markets Development Trust Fund (CMDTF), the African Financial Markets Initiative (AFMI), and the Africa Infrastructure Resilience Facility • **guarantees** to improve access to funding	**Ex-post mechanisms** • **immediate response**—emergency funds, which can be used if disaster is imminent or reoccurring • **reconstruction funding**—traditional financing, or funding for certain countries from the Fragile States Facility **Ex-ante mechanisms** • **Africa Disaster Risk Financing Program**, to increase the capacity of countries to evaluate climate risks and risk mitigation strategies, offer funding post-disaster for immediate response, and help countries access risk transfer services through "premium support" for parametric weather insurance, a product offered by the African Risk Capacity (ARC)

Knowledge sharing

- **support for the Africa Global Center on Adaptation**, including an infrastructure resilience accelerator
- **coordination mechanism, made available by the AfDB as member of the "climate finance readiness mechanism"** convened by the Green Climate Fund Secretariat, with the aim of improving coordination between the different programs and institutions providing climate change capacity/readiness support to countries
- **support for governments in data collection**, and financing offered for data generation projects, such as satellites

ADRiFi = Africa Disaster Risk Financing Program; ARC = African Risk Capacity Insurance Company Limited; AfDB = African Development Bank.
Source: Vivid Economics, based on literature reviews and stakeholder interviews.

Table A.2: Asian Development Bank Offering

Risk-informed planning
<ul><li>**technical assistance (TA) for institutional development** to mainstream climate action into national development planning; support climate investment strategies through country partnership strategies</li><li>**Climate Risk Management Framework**, defining guidance in project design to assess, target, and monitor climate risk, together with project preparation and capacity development support</li><li>**sector-wide planning support** through sector-specific resilience guidelines and plans, as well as through technical assistance in client countries</li><li>**SOURCE platform** for sharing experience and best practice in resilient infrastructure implementation</li><li>New **GIS platform for risk assessment** (currently being developed, availability expected by 2023)</li></ul>

Financing mechanisms	
Resilient infrastructure finance	**Disaster risk finance**
Instruments<ul><li>**traditional loans, concessional loans, policy based-loans, and grants**—main sources of infrastructure finance</li><li>**Asian Development Fund and climate funds** (including Urban Climate Change Resilience Trust Fund), providing targeted financing for climate-resilient infrastructure</li><li>**holistic approach to resilience, with project-financed support**, considering physical, financial, ecological, and social/institutional resilience, as stated in ADB's revised Disaster and Emergency Assistance Policy</li></ul>	**Ex-post mechanisms**<ul><li>**post-disaster needs assessment,** with TA as well as ADB staff support</li><li>**immediate response support** through emergency assistance loans, the Asia Pacific Disaster Response Fund, the Small Expenditure Financing Facility (providing capped amounts directly after a disaster event), the Disaster Response Facility (providing up to an entire yearly allocation), and mechanisms allowing projects to be refinanced or adapted</li><li>**reconstruction funding**—traditional infrastructure financing mechanisms</li><li>**"building back better,"** integrated into some mechanisms (emergency assistance loans)</li></ul>
Market development<ul><li>**national capital market development programs**, made available through loans and technical assistance, including those provided to finance ministries</li><li>**support in creating an enabling environment**</li><li>**work in collaboration with PPP agencies**, to incorporate resilience components</li></ul>	**Ex-ante mechanisms**<ul><li>**TA for designing insurance schemes**, provided to governments and offering support for future operations</li><li>**mechanisms for immediate response**, such as contingent disaster financing, providing DMCs with immediate liquidity after natural hazards, following in-depth policy dialogue on disaster preparedness and building back better</li></ul>**Market development**<ul><li>**TA for improving insurance markets**, provided to governments</li></ul>

Knowledge sharing
<ul><li>commissioned research, aimed at understanding climate change and adaptation, and making relevant data and technical resources available</li><li>training sessions, organized to ensure uptake of products</li><li>best-practice guides, to promote private sector investment in certain sectors</li><li>sector-specific guidance, offered to promote the climate-proofing of infrastructure projects</li><li>guidance on key steps involved in taking resilience into account in system-wide infrastructure planning</li><li>support for knowledge sharing between DMCs</li></ul>

ADB = Asian Development Bank; DMC = developing member country, PPP = public-private partnership.
Source: Vivid Economics, based on literature reviews and stakeholder interviews

Table A.3: Asian Infrastructure Investment Bank Offering

Risk-informed planning

- **sector-specific infrastructure resilience strategies**, including an integrated approach to sustainable city development
- **standardized environment and social framework for infrastructure projects**, including a section on systematic climate risk screening (other screening tools are currently being assessed as potential alternatives)

Financing mechanisms: Resilient infrastructure finance

Instruments

- **sovereign and nonsovereign financing**, including project preparation funding targeted at infrastructure resilience assessments
- **issuance of AIIB bonds**, with proceeds to be used for sustainable infrastructure
- **Special Fund Window (SFW)** to make AIIB financing more affordable to less-developed members
- **possibility of technical assistance** in the future

Market development

- **development of debt capital markets for infrastructure**, for example, through the Sustainable Capital Markets Initiative
- **support for the mobilization of private finance** through transactions based on third-party referrals of readily investable projects, or the origination, structuring, and execution of stand-alone deals
- **Asia Climate Bond** portfolio—fixed-income portfolio for the improvement of the climate bond market
- **equity investments**, including those made through Lightsmith Climate Resilience Partners, in growth-stage companies that develop and provide climate resilience technologies, products, and services

Knowledge sharing

- **Sustainable Cities Strategy**—project financing with a clear integrated approach, as part of a broader citywide planning strategy, giving investment priority to improving resilience against disasters, with emphasis on nature-based solutions
- **tool for use by investors in identifying climate champions**, in three measurement categories—portion of green business activities, climate mitigation, and climate resilience
- **Asian Infrastructure Finance Report**, providing insights derived from AIIB's experience

AIIB = Asain Infrastructure Investment Bank.
Source: Vivid Economics, based on literature reviews and stakeholder interviews.

Table A.4: European Bank for Reconstruction and Development Offering

Risk-informed planning

- Green Economy Transition Strategy—approach to helping countries build green, low-carbon, and resilient economies
- inclusion of resilience among performance requirements for all projects
- resilience and climate review procedure for projects
- EBRD Green Cities—financing provided to make cities more sustainable, by identifying, prioritizing, and connecting their environmental challenges with sustainable infrastructure investments and policy measures
- discussions with GCF and the Global Center on Adaptation on how to improve risk management at the network scale

Financing mechanisms: Resilient infrastructure finance

Instruments

- **loans, equity investments, and cofinancing**
- **technical cooperation grants**
- **green bonds**, including a climate resilience bond

Market development

- **publication of national market development strategy reports**
- **Local Currency and Capital Markets Development Initiative**, to promote the use of local currencies and the development of local markets, through technical cooperation, loans, regulatory support, and monitoring
- **SME Local Currency Programme**, involving funding and the promotion of policy dialogue to improve local markets and local lending, and ensure that companies can access affordable funding
- **private sector investment promotion** through the creation of investment-friendly regulatory frameworks or facilities, such as the Finance and Technology Transfer Centre for Climate Change, which aims to facilitate the use of climate technologies by private companies
- **guarantees** for private sector investment
- **support for financial institutions** in assessing climate action and incorporating it into their portfolios
- **collaboration with the Global Center on Adaptation** in incorporating resilience into PPPs

Knowledge sharing

- **production of best-practice handbook, incorporating climate risk considerations in the financial sector; training for corporate banks; and engagement with the Climate Action in Financial Institutions Initiative**
- **presentation of the impact of resilience in the EBRD Sustainability Report**, according to six outcome categories

EBRD = European Bank for Reconstruction and Development, GCF = Green Climate Fund, PPP = public-private partnership, SME = small and medium-sized enterprises.
Source: Vivid Economics, based on literature reviews and stakeholder interviews.

Table A.5: European Investment Bank Offering

Risk-informed planning

- **climate risk assessment framework** for projects, with physical, economic, and environmental risk dimensions
- **advisory services** on market or project development and on climate adaptation at every step of resilience project development, to improve adaptive capability

Financing mechanisms: Resilient infrastructure finance

Instruments

- **loans and equity investments**

Market development

- **sector studies** that highlight market gaps, which can be addressed with the help of financing mechanisms described above
- **co-implementation of funds**, encouragement of private finance
- **guarantees** for private sector investments

Knowledge sharing

- **support for improvements in international sustainable finance**, such as harmonizing standards and definitions for sustainable investing
- **co-creation of depositories of good practice**

Source: Vivid Economics, from literature review and stakeholder interviews.

Table A.6: Inter-American Development Bank Offering

Risk-informed planning

- **disaster risk profiles** published at country level, for sector/area/hazard prioritization (strategic country risk evaluation)
- **indicators of disaster risk and risk management** for countries to use
- **technical support for project risk assessments**, with TA support
- **policy support and assistance**, to improve disaster risk management and incorporate climate risk into infrastructure plans
- **sustainable infrastructure framework**—disaster and climate risk assessment methodology for all projects
- **standardized disaster and climate risk assessment** for projects, and Blue Spot analysis conducted where relevant

Financing mechanisms

Resilient infrastructure finance	Disaster risk finance
Instruments	**Ex-post mechanisms**
- **loans (project and policy-based) and equity investments** - **focus on nature-based solutions**, but no dedicated instruments	- **post-disaster needs assessment**, supported through the Immediate Response Facility for Emergencies Caused by Disasters - **immediate funding**, offered through the Immediate Response Facility - **reconstruction funding**, offered through the usual financing mechanisms, and loan reformulation
Market development	**Ex-ante mechanisms**
- **improvements in readiness for accessing climate finance** through improvements in government capacity for project development, and support for funding access - **guarantees** for private sector investments - **improvements in green bond reporting** through the Green Bond Transparency Platform	- **Caribbean Catastrophe Risk Insurance Facility**—coverage modality based on parametric modeling of exposed population - **immediate funding**, offered through two contingent credit facilities
	Market development
	- **support for the design and implementation of CAT bonds**

Knowledge sharing

- **Index of Governance and Public Policy in Disaster Risk Management**, for evaluating governmental risk management capacity, and training of stakeholders and clients in its use
- **training of staff in nature-based solutions**, the challenges in delivering them, and the project stages at which they are most relevant
- **networks of policy-makers** in disaster risk management and climate change (Disaster Risk Management Network, Climate Change Network)
- **knowledge transfer and capacity building through online courses**, including DRM courses for bank specialists and client agencies, and open as well as private courses in integrating disaster and climate change risk considerations into infrastructure projects

CAT bonds = catastrophe bonds, DRM = disaster risk management, TA = technical assistance.
Source: Vivid Economics, based on literature reviews and stakeholder interviews.

Table A.7: Islamic Development Bank Offering

Risk-informed planning

- **support for the development of risk models**
- **promotion of the use of early warning systems**
- **climate-screening of projects** using the World Bank tool
- **promotion of strategic climate action** through inclusion in country partnership strategies
- **institutional capacity building** (including focus on inter-sector linkages)
- **improvement of national risk governance**

Financing mechanisms

Resilient infrastructure finance	Disaster risk finance
Instruments - **loans and equity investments** **Market development** - **promotion of development of Islamic financial and capital markets**	**Ex-post mechanisms** - **support for post-disaster needs assessment** through capacity building - **reconstruction funding**, offered through traditional financing mechanisms, with the objective of building back better, as well as strengthening institutional capacity to respond to disasters - **support for South–South cooperation** in building assistance networks for emergencies - **building back better**, as stated focus of post-disaster finance - **contingent response component** typically included in projects, allowing for funding reallocation in case of a disaster

Knowledge sharing

- **contributions to global knowledge and legislation on climate change**
- **sector-wide climate adaptation guidance notes**
- **contributions to making early warning systems more accessible and institutionalized**

Source: Vivid Economics, based on literature reviews and stakeholder interviews.

Table A.8: World Bank Offering

Risk-informed planning

- **risk assessment**—National-/Policy-Level Climate and Disaster Risk Screening Tool (used in systematic country diagnostics), Risk Stress Testing Tool (RiST) (for integrating risk assessment into economic assessment of projects), Open Data for Resilience Initiative, standard tools for vulnerability assessments and recommendations
- **resilience rating system** for all World Bank or other MDB projects (process-focused)
- **climate change and disaster risk screening** (systematic evaluation of national-level climate risk) for all new projects
- **sector- and system-level capacity support** through the City Resilience Program
- **Global Facility for Disaster Reduction and Recovery (GFDRR)**—global partnership established to help developing countries understand better and reduce their vulnerability to natural hazards and climate change

Financing mechanisms

Resilient infrastructure finance	Disaster risk finance
Instruments	**Ex-post mechanisms**
• **traditional loans, concessional loans, and grants**—main sources of infrastructure finance • **Program-for-Results financing**, which can be targeted at disaster preparedness of infrastructure • **green bonds** issued by the World Bank, with proceeds intended for the financing of infrastructure disaster preparedness	• **post-disaster needs assessment**, offered by the GFDRR Standby Recovery Financing Facility • **immediate response**—development policy loans with a catastrophe deferred drawdown option, community-driven development (emergency lending to sub-sovereign entities after disasters), and Crisis Response Window (emergency financing of last resort) • **reconstruction funding**, supported through loan and grant mobilization • **building-back-better benefits**, targeted across DRF mechanisms, but often without explicit criteria for funding access
Market development	**Ex-ante mechanisms**
• **guarantees**, to crowd in private sector investment (policy-based or through MIGA) • **capital market development**—financing for financial reform, advisory support for the use of capital market instruments in infrastructure financing • **coordination**—training and support for private investors, stakeholders, and policy-makers to facilitate bond market development, institutional investor involvement, and creation of PPP frameworks (ESMID, Deep Dive) • **networks** connecting stakeholders and providing quality assurance for experts from the network	• **solutions for financial recovery**, provided through the Disaster Risk Insurance Platform (offering funding and expertise) • **immediate response**, offered through contingent emergency response components of infrastructure investments
	Market development
	• **transfer of catastrophe risk to the private reinsurance market**, supported through risk transfer Instruments, and multi-country catastrophe risk pools • **support in designing public asset insurance mechanism**s—Southeast Asia Disaster Risk Insurance Facility (SEADRIF)

Knowledge sharing

- **global community of experts and practitioners in risk assessment and communication**—Understanding Risk community of practice
- **open-source geospatial data tools for risk assessments**

DRF = Disaster Response Facility, ESMID = Efficient Securities Markets Institutional Development, MDB = multilateral development bank; MIGA = Multilateral Investment Guarantee Agency, PPP = public–private partnership.

Source: Vivid Economics, based on literature reviews and stakeholder interviews.

References

African Development Bank (AfDB). 2008. *Revised Policy Guidelines and Procedures for Emergency Relief Assistance*. Abidjan, Côte d'Ivoire.

————. 2009. *Bank Group Climate Risk Management and Adaptation Strategy (CRMA)*. Abidjan, Côte d'Ivoire.

————. 2011. *AfDB's Financial Products*. Abidjan, Côte d'Ivoire.

————. 2013. *African Development Bank Group's Integrated Safeguards System: Policy Statement and Operational Safeguards*. Safeguards and Sustainability Series. Vol. 1, Issue 1. December. Abidjan, Côte d'Ivoire.

————. 2016. Africa Thriving and Resilient: The African Development Bank Group's Second Climate Change Action Plan (2016–2020). Abidjan, Côte d'Ivoire.

————. 2018. African Development Bank Rolls Out Programme to Boost Climate Risk Financing and Insurance for African Countries. News release.

————. 2020a. *Accelerating Africa's Climate Resilient and Low-Carbon Development: 2019 Annual Report of the AfDB Climate Change and Green Growth Department*. Abidjan, Côte d'Ivoire.

————. 2020b. *2019 Annual Report*. Abidjan, Côte d'Ivoire.

————. 2021a. *Annual Report 2020*. Abidjan, Côte d'Ivoire.

————. 2021b. *Policy on Water*. Abidjan, Côte d'Ivoire.

————. 2022. *Climate Safeguards System (CSS): Climate Screening and Adaptation Review & Evaluation Procedures Booklet*. Abidjan, Côte d'Ivoire.

————. n.d.[a]. Adaptation Benefit Mechanism (accessed January 2022).

————. n.d.[b]. Africa Disaster Risks Financing (ADRiFi) Programme.

————. n.d.[c]. African Financial Markets Initiative (AFMI) (accessed August 2021).

————. n.d.[d] Capital Markets Development Trust Fund (CMDTF) (accessed August 2021).

————. n.d.[e]. ClimDev Special Fund (accessed August 2021).

————. n.d.[f]. *Environmental & Social Assessment Procedures: Basics*.

————. n.d.[g] Portfolio Selection (accessed August 2021).

————. n.d.[h]. Projects & Operations (accessed January 2022).

AfDB, Asian Development Bank (ADB), Asian Infrastructure Investment Bank (AIIB), European Bank for Reconstruction and Development (EBRD), European Investment Bank (EIB), Inter-American Development Bank (IDB), International Development Finance Club (IDFC), and Islamic Development Bank (IsDB). 2019. *A Framework and Principles for Climate Resilience Metrics in Financing Operations*. Washington, DC: IDB.

AfDB, ADB, AIIB, EBRD, EIB, IDB, IsDB, New Development Bank (NDB), and World Bank. 2021a. _Common Principles for Climate Mitigation Finance Tracking_. Version 3. 18 October. Frankfurt, Germany: IDFC.

———. 2021b. _2020 Joint Report on Multilateral Development Banks' Climate Finance_. London: EBRD.

AfDB, ADB, EBRD, EIB, IDB, World Bank, and International Finance Corporation (IFC). 2012. _Joint MDB Report on Mitigation Finance 2011_. Washington, DC: World Bank.

African Risk Capacity Insurance Company Limited. 2021. Settling a Claim in Madagascar Proves the Value of Insurance. 13 December.

Antonich, B. 2020. UN Estimates the Global Cost of Adaptation. 14 February. _Global Center on Adaptation_.

Ashwill, M., and L. Alvarez. 2014. _Climate Change and IDB: Building Resilience and Reducing Emissions_. Sector study done for the IDB evaluation project. Washington, DC: IDB.

ADB. 2001. _Developing Best Practices for Promoting Private Sector Investment in Infrastructure: Roads_. Manila.

———. 2009a. _Establishment of the Asia Pacific Disaster Response Fund_. Manila.

———. 2009b. _Safeguard Policy Statement_. Manila.

———. 2011. _Summary of ADB Financial Instruments and Approval Procedures_. Manila.

———. 2013a. _Investing in Resilience: Ensuring a Disaster-Resistant Future_. Manila.

———. 2013b. _Urban Operational Plan 2012–2020_. Manila.

———. 2014. _Climate Risk Management in ADB Projects_. Manila.

———. 2015a. Frequently Asked Questions: Enhancing ADB's Financial Capacity by Up to 50% for Reducing Poverty in Asia and the Pacific: Combining ADB's ADF OCR Resources. 31 March. Manila.

———. 2015b. _Jiangxi Pingxiang Integrated Rural–Urban Infrastructure Development: Final Report_. Consultant's report. Manila (TA 8451-PRC).

———. 2017a. _Climate Change Operational Framework 2017–2030_. Manila.

———. 2017b. _Disaster Risk Assessment for Project Preparation: A Quick Guide_. Manila.

———. 2017c. _Disaster Risk Management and Country Partnership Strategies: A Practical Guide_. Manila.

———. 2017d. _Meeting Asia's Infrastructure Needs_. Manila.

———. 2019f. Small Expenditure Financing Facility. Operations Manual Policies and Procedures. Manila.

———. 2018a. _PCDIP: Philippine City Disaster Insurance Pool_. Manila.

———. 2018b. _Strategy 2030: Achieving a Prosperous, Inclusive, Resilient, and Sustainable Asia and the Pacific_. Manila.

———. 2018c. Technical Assistance Completion Report: Enhancing Insurance Market Efficiency and Outreach in Kazakhstan. Manila.

———. 2019a. *Expanded Disaster Response Facility under ADF 13*. Paper prepared for the Asian Development Fund (ADF) 13 Replenishment Meeting. Manila. 5–7 November.

———. 2019b. *Strategy 2030 Operational Plan for Priority 4: Making Cities More Livable, 2019–2024*. Manila.

———. 2019c. Technical Assistance Completion Report: *Support for Post Typhoon Yolanda Disaster Needs Assessment and Response in the Philippines*. Manila.

———. 2020a. *Asian Development Fund 13 Donors' Report: Tackling the COVID-19 Pandemic and Building a Sustainable and Inclusive Recovery in Line with Strategy 2030*. Manila.

———. 2020b. *Budget of the Asian Development Bank for 2021*. Manila.

———. 2021a. ADB to Enhance Adaptation and Resilience in Response to Climate Change Threat. 4 May.

———. 2021b. Energy Policy: Supporting Low Carbon Transition in Asia and the Pacific. Draft policy paper.

———. 2022c. Policy-Based Lending. Operations Manual Policies and Procedures. Manila.

———. 2021d. Revised Disaster and Emergency Assistance Policy. Draft policy paper.

———. 2021e. *A System-Wide Approach for Infrastructure Resilience: Technical Note*. Manila.

———. 2022. *Disaster-Resilient Infrastructure: Unlocking Opportunities for Asia and the Pacific*. Manila.

———. n.d.[a]. ADB's Work in Sustainable Transport (accessed August 2021).

———. n.d.[b]. Asian Development Fund 13 (accessed August 2021).

———. n.d.[c]. Funds and Resources. Climate Change Financing at ADB (accessed August 2021).

———. n.d.[d]. Climate Change Fund (accessed August 2021).

———. n.d.[e]. Funds and Resources. Project Readiness Improvement Trust Fund (accessed August 2021).

———. n.d.[f]. Projects and Tenders (accessed January 2022).

———. n.d.[g]. Publications and Documents: Climate Risk Country Profiles (accessed January 2022).

———. n.d.[h]. Regional: Pacific Disaster Resilience Program (accessed January 2022).

———. n.d.[i]. Safeguard Policy Review (accessed August 2021).

———. n.d.[j]. Funds and Resources. Technical Assistance Special Fund.

———. n.d.[k]. Tonga and ADB (accessed January 2022).

ADB, AfDB, AIIB, EIB, EBRD, IDB, IDB Invest, Islamic Corporation for the Development of the Private Sector (ICD), IsDB, and World Bank. 2018. *Global Toolbox: Instruments Available from Multilateral Development Banks to Support Private Investment in Asia*.

AIIB. 2018a. **Strategy on Mobilizing Private Capital for Infrastructure**. 9 February. Beijing.

———.2019. AIIB and Amundi Launch Innovative USD500-Million Climate Bond Portfolio to Mobilize Climate Action. 10 September. London and Beijing.

————. 2020a. AIIB Debut in GBP Market Unlocks New Financing for Sustainable Development. 14 October. London.

————. 2020b. *Corporate Strategy: Financing Infrastructure for Tomorrow*. Beijing.

————. 2020d. *Water Sector Strategy*. Beijing.

————. 2021a. AIIB Joins Global Infrastructure Facility to Enhance Multilateral Cooperation. Beijing. 8 July.

————. 2021b. *Environmental and Social Framework*. Beijing.

————. 2021c. *Sustainable Development Bond Framework*. Beijing.

————. 2021d. *2020 AIIB Annual Report*. Beijing.

————. 2021 Business Plan and Budget Summary. Unpublished.

————. n.d.[a]. Financing Operations (accessed August 2021).

————. n.d.[b]. Multicountry: Asia Climate Bond Portfolio (accessed August 2021).

————. n.d.[c]. Our Investments: 2019 AIIB Annual Report and Financials (accessed August 2021).

————. n.d.[d]. Our Projects (accessed January 2022).

————. n.d.[e]. Project Preparation Special Fund (PPSF) (accessed August 2021).

————. Sustainable Cities Strategy: Financing Solutions for Developing Sustainable Cities in Asia. Unpublished.

————. n.d.[f]. Transport Sector Strategy: Sustainable and Integrated Transport for Trade and Economic Growth in Asia.

————. n.d.[g]. Who We Are: Infrastructure for Tomorrow (accessed August 2021).

Barandiarán, M. et al. 2019. *Disaster and Climate Risk Assessment Methodology for IDB Projects: A Technical Reference Document for IDB Project Teams*. Technical Note No. TN-01771. Washington, DC.: IDB.

Bennett, V. 2019. World's First Dedicated Climate Resilience Bond, for US$700m, Is Issued by EBRD. *EBRD*. 20 September.

City Climate Finance Gap Fund (Gap Fund). n.d. Turn Resilient Low-Carbon Ideas into Strategies and Finance-Ready Projects (accessed August 2021).

Climate Bonds Initiative (CBI). 2018. *Why Making Infrastructure Climate-Adapted and Resilient Will Help Meet the SDGs*. Briefing paper. London.

————. n.d. Climate Resilience Principles (accessed January 2022).

Climate Policy Initiative. 2018. *Implementing the EBRD Green Economy Transition*. A summary of the EBRD Green Economy Transition manual. London.

Coalition for Climate Resilient Investment (CCRI). n.d. Physical Climate Risk Assessment Methodology.

Coppola, D. P. 2015. *Introduction to international Disaster Management*. 3rd ed. Elsevier Science.

CRiSTAL. n.d. CRiSTALTool.org: Community-based Risk Screening Tool – Adaptation and Livelihoods (accessed January 2022).

Deschryver, Pauline, and Frederic de Mariz. 2020. What Future for the Green Bond Market? How Can Policymakers, Companies, and Investors Unlock the Potential of the Green Bond Market? *Journal of Risk and Financial Management*. 13 (3). 61.

Dickson, E. et al. 2022. *Urban Risk Assessments: Understanding Disaster and Climate Risk in Cities*. Urban Development Series. Washington, DC: World Bank.

Dwyer, G. 2017. Piloting "Sponge Cities" in the People's Republic of China. Project result/Case study. *ADB*. Manila. 13 January.

edX. n.d. Natural Disasters and Climate Change Risk Assessment in Infrastructure Projects (accessed April 2022).

Energy Sector Management Assistance Program (ESMAP). 2010. *HEAT: Hands-On Energy Adaptation Toolkit*. Washington, DC: World Bank.

Environmental XPRT. n.d. Acclimatise Aware – Online, Rapid, Climate Risk Screening Tool (accessed January 2022).

EBRD. 2013. Guide to EBRD Financing.

———. 2018a. *Energy Sector Strategy (2019–2023)*. London.

———. 2018b. *LC2 Strategy 2019–2024*. As approved by the EBRD Board of Directors on 28 November, for Local Currency and Capital Markets Development (LC2). London.

———. 2018c. Trade Facilitation Programme. London.

———. 2019a. *Environmental and Social Policy*. London.

———. 2019b. *Municipal and Environmental Infrastructure Sector Strategy*. London.

———. 2019c. *Transport Sector Strategy (2019–2024)*. London.

———. 2020. *Green Economy Transition Approach 2021–2025*. London.

———. 2021a. *Sustainability Report 2020*. London.

———. 2021b. *Task Force on Climate-Related Financial Disclosures Report 2020*. London.

———. n.d.[b]. *Building Resilience to Climate Change: Investing in Adaptation*.

———. n.d.[c]. EBRD's Green Bond Issuance (accessed August 2021).

———. n.d.[d]. The EBRD's Local Currency and Capital Markets Development Initiative (accessed August 2021).

———. n.d.[e]. EBRD SME Local Currency Programme (accessed August 2021).

———. n.d.[f]. Effective Policy Instruments for Green Cities (accessed August 2021).

———. n.d.[g]. Loans (accessed August 2021).

———. n.d.[h]. Project Finder (accessed January 2022).

————. n.d.[i]. Supporting Tools and Toolkit. *EBRD Environmental and Social Risk Management Manual (E-Manual)* (accessed January 2022).

————. n.d.[j]. What FINTECC and ENVITECC Offer (accessed August 2021).

————. n.d.[k]. What Is the EBRD's Green Economy Transition Approach? (accessed August 2021).

————. n.d.[l]. What We Do in Municipal Infrastructure (accessed August 2021).

EBRD Green Cities. n.d. About Green Cities (accessed August 2021).

EIB. 2018. *Environmental and Social Standards*. Luxembourg.

————. 2020a. The EIB and UNECE Establish an Information Repository of Good Practices and Lessons Learned in Land-Use Planning and Industrial Safety. 16 December.

————. 2020b. *EIB Climate Strategy*. Luxembourg.

————. 2020c. *EIB Group Climate Bank Roadmap 2021–2025*. Luxembourg.

————. 2020d. Post Disaster Reconstruction & Prevention: Spain.

————. 2020e. Sustainable Transport: Overview.

————. 2021a. Climate Action and Environmental Sustainability: Overview.

————. 2021b. Energy Overview 2021.

————. 2021c. Madagascar: EIB Accelerating Climate Disaster Recovery, Priority Clean Energy and Transport Investment. 14 January.

————. 2021d. MDBs' Climate Finance Rose to US$66 Billion in 2020, Joint Report Shows. 21 June.

————. n.d.[a]. Climate and Environmental Sustainability (accessed August 2021).

————. n.d.[b]. Climate Awareness Bonds (accessed August 2021).

————. n.d.[c]. EIB at a Glance (accessed January 2022).

————. n.d.[d]. Market Development (accessed August 2021).

————. n.d.[e]. Projects (accessed January 2022).

————. n.d.[f]. Water and Wastewater Management (accessed August 2021).

————. n.d.[g]. What We Offer (accessed August 2021).

Giuliani, G. and P. Peduzzi. 2011. The PREVIEW Global Risk Data Platform: A Geoportal to Serve and Share Global Data on Risk to Natural Hazards. *Natural Hazards and Earth System Sciences*. 11. pp. 53–66.

Global Center on Adaptation (GCA). n.d. Africa Adaptation Acceleration Program (accessed August 2021).

Global Facility for Disaster Reduction and Recovery (GFDRR). 2014. *Understanding Disaster Risk in an Evolving World: A Policy Note*. Washington, DC: World Bank.

———. 2021. *GFDRR Strategy 2021–2025: Scaling Up and Mainstreaming Resilience in a World of Compound Risks*. Washington, DC: World Bank.

———. n.d.[a]. Funding Structure and Partnerships (accessed August 2021).

———. n.d.[b]. Global Facility For Disaster Reduction and Recovery (accessed August 2021).

———. n.d.[c]. Open Data for Resilience Initiative (OpenDRI) Background.

Green Climate Fund (GCF). n.d. Asian Development Bank (accessed August 2021).

Guerrero Compeán, R., and S. Lacambra Ayuso. 2020. Disasters and Loss of Life: New Evidence on the Effect of Disaster Risk Management Governance in Latin America and the Caribbean. *IDB Working Paper Series.* No. IDB-WP-01126. Washington, DC: IDB.

Hallegatte, S. et al. 2021. *Integrating Climate Change and Natural Disasters in the Economic Analysis of Projects: A Disaster and Climate Risk Stress Test Methodology*. Washington, DC: World Bank.

IDB Invest. n.d. About Us (accessed August 2021).

Infrastructure Outlook. n.d. Forecasting Infrastructure Investment Needs and Gaps (accessed January 2022).

Inter-American Development Bank (IDB). 2010. *Indicators of Disaster Risk and Risk Management: Program for Latin America and the Caribbean Summary Report*. Washington, DC.

———. 2013. *Mid-term Evaluation of IDB-9 Commitments: Lending Instruments*. Background paper, prepared by the Office of Evaluation and Oversight (OVE). Washington, DC.

———. 2013. Peru Improves Ability to Respond to Natural Disasters with IDB Support. 13 December. Washington, DC.

———. 2018a. *Public Sector Financing: Lending Instruments and Guarantees*. Washington, DC.

———. 2018b. *What Is Sustainable Infrastructure? A Framework to Guide Sustainability Across the Project Cycle*. Technical Note No. IDB-TN-1388. Washington, DC.

———. 2020. *Sustainability Report 2019*. Washington, DC.

———. 2021a. *Annual Report 2020: The Year in Review*. Washington, DC.

———. 2021b. IDB Announces Build Forward Initiative to Advance Technology and Resilience in the Caribbean. 25 February.

———. 2021c. *Inter-American Development Bank Group Climate Change Action Plan 2021–2025*. Washington, DC.

———. n.d.[a]. Climate Change Network (accessed August 2021).

———. n.d.[b]. Concessional Resources (accessed August 2021).

———. n.d.[c]. Disaster Risk Management (accessed August 2021).

———. n.d.[d]. Disaster Risk Management Network (accessed January 2022).

———. n.d.[e]. Environment and Natural Disasters: Creating Sustainable Development Opportunities and Safeguards (accessed January 2022).

———. n.d.[f]. Financial Markets (accessed August 2021).

———. n.d.[g]. Helping Latin America and the Caribbean Manage Natural Disaster Risk (accessed August 2021).

———. n.d.[h]. IDB Projects (accessed January 2021).

———. n.d.[i]. Investments (accessed August 2021).

———. n.d.[j]. Natural Disasters: Preparedness Response, and Recovery (accessed August 2021).

———. n.d.[k]. *The New IDB Insurance Facility: Natural Disasters Insurance Facility for Central America and the Caribbean*.

———. n.d.[l]. Regional Policy Dialogue: Disaster Risk Management Network (accessed August 2021).

———. n.d.[m]. Special Development Lending Category (accessed August 2021).

———. n.d.[n]. Technical Cooperation (accessed August 2021).

———. n.d.[o]. Transportation in Latin America and the Caribbean (accessed August 2021).

———. n.d.[p]. Water and Sanitation: Clean Waters for Latin America and the Caribbean (accessed August 2021).

Intergovernmental Panel on Climate Change (IPCC). 2021. *Climate Change 2021: The Physical Science Basis*. Geneva.

International Aid Transparency Initiative (IATI). n.d. Units of Aid (accessed August 2021).

International Development Association (IDA). n.d.[a]. Crisis Response Window (accessed August 2021).

———. n.d.[b]. Immediate Response Mechanism (accessed August 2021).

IFC. n.d. Climate Risk and Adaptation (accessed August 2021).

International Monetary Fund (IMF). n.d. SDR per Currency Unit and Currency Units per SDR Last Five Days (accessed August 2021).

Islamic Corporation for the Development of the Private Sector (ICD). n.d. About ICD (accessed August 2021).

IsDB. 2005. *Islamic Development Bank Group In Brief*. Jeddah, Saudi Arabia.

———. 2007. *Islamic Development Bank Group in Brief*. Jeddah, Saudi Arabia.

———. 2019. *Agriculture & Rural Development Sectors Climate Change Adaptation: Guidance Note*. Jeddah, Saudi Arabia.

———. 2019a. *Climate Change Policy*. Jeddah, Saudi Arabia.

———. 2019b. *Disbursement Handbook 2019*. Jeddah, Saudi Arabia.

———. 2019c. *Energy Sector Climate Change Adaptation: Guidance Note*. Jeddah, Saudi Arabia.

———. 2020a. *Disaster Risk Management and Resilience Policy*. Jeddah, Saudi Arabia.

———. 2018b. *Reverse Linkage: Development through South–South Cooperation*. Jeddah, Saudi Arabia.

———. 2020c. *2019 Annual Report: Shaping New Frontiers for Sustainable Development*. Jeddah, Saudi Arabia.

———. 2020d. *2020–2025 Climate Action Plan*. Jeddah, Saudi Arabia.

———. 2021. Islamic Development Bank Issues Largest Sustainability Sukuk Ever. 25 March.

———. n.d. Projects Database (accessed January 2022).

Jha, A. K., and Z. Stanton-Geddes, eds. 2013. *Strong, Safe, and Resilient: A Strategic Policy Guide for Disaster Risk Management in East Asia and the Pacific*. Directions in Development. Washington, DC: World Bank.

Karpinski, K. 2019. *Philippines Financial Sector Assessment Program: Retail Payments*. Technical note. Washington, DC: World Bank.

Kornejew, M., J. Rentschler, and S. Hallegatte. 2019. Well Spent: How Governance Determines the Effectiveness of Infrastructure Investments. *Policy Research Working Paper*. No. 8894. Washington, DC: World Bank.

Lacambra, S. et al. 2015a. *Index of Governance and Public Policy in Disaster Risk Management (iGOPP): Application Protocol*. Washington, DC: IDB.

———. 2015b. *Index of Governance and Public Policy in Disaster Risk Management (iGOPP): Main Technical Document*. Washington, DC: IDB.

Lacambra, Sergio et al. 2011. *Disaster Risk Management and Long-Term Adaptation Approach at the Inter-American Development Bank (IDB): Synthesis Report*. Washington, DC: IDB.

Lawrance, C. 2019. The Rise of Sustainable Capital Markets. *AIIB*. 12 July.

Lu, X. 2019. Building Resilient Infrastructure for the Future: Background Paper for the G20 Climate Sustainability Working Group. *ADB Sustainable Development Working Paper Series*. No. 61. Manila: Asian Development Bank.

Mesoamerican Reef Fund, InsuResilience Solutions Fund, and Willis Towers Watson. 2021. Innovative Post-Hurricane Protection for Endangered Mesoamerican Coral Reef Goes Live with Insurance Carrier Confirmed. 30 July.

Milman, O. et al. 2021. The Climate Disaster Is Here. *The Guardian*. 14 October.

Oh, H. E. et al. 2019. *Addressing Climate Change in Transport: Vol 2: Pathway to Resilient Transport*. Vietnam Transport Knowledge Series. Washington, DC: World Bank.

Organisation for Economic Co-operation and Development (OECD). 2014. Islamic Development Bank Statistical Reporting to the OECD Development Assistance Committee (DAC). Note prepared by the Statistics and Development Finance Division of the OECD Development Co-operation Directorate. March.

Panwar, V. n.d. *Financing Resilient Infrastructure: Bridging the Funding Gap*. Blog. New Delhi: Coalition for Disaster Resilient Infrastructure (CDRI).

Rau, S. 2021. Sponge Cities in the People's Republic of China—Evolution of ADB Support. Presentation at the Asian Development Bank Knowledge Events webinar Nature-based Solutions for Resilient Infrastructure. 15 July.

Ray, P. A., and C. M. Brown. 2015. *Confronting Climate Uncertainty in Water Resources Planning and Project Design: The Decision Tree Framework*. Washington, DC: World Bank.

The Resilience Shift. n.d. Resilience Toolbox (accessed August 2021).

Rozenberg, J., and M. Fay. 2019. *Beyond the Gap: How Countries Can Afford the Infrastructure They Need while Protecting the Planet*. Washington, DC: World Bank.

Sirivunnabood, P., and W. Alwarritzi. 2020. Incorporating a Disaster Risk Financing and Insurance Framework into Country Management and Development Strategies. *Policy Brief*. No. 2020-5. Tokyo: Asian Development Bank Institute.

Tall, A. et al. 2021. *Enabling Private Investment in Climate Adaptation and Resilience: Current Status, Barriers to Investment and Blueprint for Action*. Washington, DC: World Bank (values for 2017–2018).

Tanner, T. et al. 2015. *The Triple Dividend of Resilience: Realizing Development Goals through the Multiple Benefits of Disaster Risk Management*. GFDRR at the World Bank, and Overseas Development Institute (ODI), London.

Tran, M., and M. Boyland. 2022. The State of Knowledge on Disaster Risk. *IRDR Working Paper Series*. Beijing: Integrated Research on Disaster Risk (IRDR).

United Nations Department of Economic and Social Affairs (UNDESA). n.d. Goal 9: Build Resilient Infrastructure, Promote Inclusive and Sustainable Industrialization and Foster Innovation.

United Nations Environment Programme / United Nations International Strategy for Disaster Reduction (UNEP/ UNISDR). n.d. PREVIEW Global Risk Data Platform (accessed January 2022).

United Nations Framework Convention on Climate Change (UNFCCC). n.d. COP26 Outcomes: Finance for Climate Adaptation (accessed December 2021).

UNFCCC 26th Climate Change Conference, Glasgow, Scotland (COP26). 2021a. *Climate Change Delivery Plan: Meeting the US$100 Billion Goal*. Bonn.

———. 2021b. MDB Joint Climate Statement. 3 November.

———. 2021c. *Progress Report: Multilateral Development Banks Working Together for Paris Alignment*. 3 November.

United Nations Office for Disaster Risk Reduction (UNDRR). 2015. *Sendai Framework for Disaster Risk Reduction 2015–2030*. Geneva.

Weichselgartner, Juergen, and Patrick Pigeon. 2015. The Role of Knowledge in Disaster Risk Reduction. *International Journal of Disaster Risk Science*. 6. pp. 107–116.

Woetzel, Jonathan, Dickon Pinner, Hamid Samandari, Hauke Engel, Mekala Krishnan, Brodie Boland, and Carter Powis. 2020. *Climate Risk and Response: Physical Hazards and Socioeconomic Impacts*. New York: McKinsey Global Institute.

World Bank. 2009. Including Contingent Emergency Response Components in Standard Investment Projects: Guidance Note to Staff. Operations Policy and Country Services.

———. 2013. *Building Resilience: Integrating Climate and Disaster Risk into Development*. Washington, DC.

———. 2015. *Investing in Urban Resilience: Protecting and Promoting Development in a Changing World*. Washington, DC.

———. 2016b. *The World Bank Group's Support to Capital Market Development*. Washington, DC.

———. 2017. The CityStrength Diagnostic: Promoting Urban Resilience. Brief. 17 October.

———. 2018. Product Note: IBRD Catastrophe Deferred Drawdown Option (Cat DDO). October. Washington, DC.

———. 2018a. Product Note: IDA Catastrophe Deferred Drawdown (Cat DDO). October. Washington, DC.

————. 2018b. *Project Appraisal Report: Anhui Road Resilience Program for Results Project in the People's Republic of China*. Washington, DC.

————. 2019a. *Green Bond Impact Report 2019*. Washington, DC.

————. 2019b. Resilient Cities. Brief. 7 October.

————. 2019c. *The World Bank Group's Action Plan on Climate Change Adaptation and Resilience : Managing Risks for a More Resilient Future*. Washington, DC.

————. 2020a. City Resilience Program. Brief. 21 December.

————. 2020b. *Disaster Risk Insurance Platform: Insurance Solutions for World Bank Clients*. Washington, DC.

————. 2021a. *Climate Change Action Plan 2021–2025: Supporting Green, Resilient and Inclusive Development*. Washington, DC.

————. 2021b. *Resilience Rating System: A Methodology for Building and Tracking Resilience to Climate Change*. Washington, DC.

————. 2021c. What You Need to Know About the Climate Change Resilience Rating System. An interview with Stéphane Hallegatte, World Bank climate change lead economist.

————. n.d.[a]. Climate & Disaster Risk Screening Tools: About the Tools (accessed January 2022).

————. n.d.[b]. Climate & Disaster Risk Screening Tools: Complementary Risk Analysis Tools and Guidance (accessed January 2021).

————. n.d.[c]. Disaster Risk Financing and Insurance (DRFI) Program (accessed August 2021).

————. n.d.[d]. IBRD Financial Products: Disaster Risk Management (accessed August 2021).

————. n.d.[e]. Investment Project Financing (IPF) (accessed August 2021).

————. n.d.[f]. National/Policy Level Climate & Disaster Screening Tool (accessed August 2021).

————. n.d.[g]. National/Policy Level Tool: Climate and Disaster Risk Screening Guidance Note (accessed January 2021).

————. n.d.[h]. Products and Services: Financing Instruments (accessed August 2021).

————. n.d.[i]. Program-for-Results Financing (PforR) (accessed August 2021).

————. n.d.[j]. Projects & Operations (accessed January 2022).

————. n.d.[k]. Topics (accessed January 2022).

————. n.d.[l]. World Bank Climate and Disaster Risk Screening Tools (accessed January 2022).

World Bank and GFDRR. 2021. *City Resilience Program Annual Report July 2019–June 2020*. Washington, DC.

Zgheib, Nibal. 2019. EBRD Boosts Capital Market Development in Jordan. *EBRD*. 10 September.

Multilateral Development Bank Support for Disaster-Resilient Infrastructure Systems

This publication explores how multilateral development banks (MDBs) can help improve infrastructure resilience in light of increasing climate and disaster risks. It highlights opportunities in three areas: (i) risk-informed planning; (ii) financing assistance that enables improvements in infrastructure resilience before, during, and after a disaster; and (iii) knowledge building through regional and global networks. Examples from around the world showcase some of the successes that have been achieved so far. The publication draws on a literature review, insights from stakeholders in developing member countries of the Asian Development Bank, and input from specialists in eight MDBs.

About the Asian Development Bank

ADB is committed to achieving a prosperous, inclusive, resilient, and sustainable Asia and the Pacific, while sustaining its efforts to eradicate extreme poverty. Established in 1966, it is owned by 68 members —49 from the region. Its main instruments for helping its developing member countries are policy dialogue, loans, equity investments, guarantees, grants, and technical assistance.

ISBN 978-92-9269-682-5

ADB

ASIAN DEVELOPMENT BANK
6 ADB Avenue, Mandaluyong City
1550 Metro Manila. Philippines
www.adb.org

9 789292 696825

www.ingramcontent.com/pod-product-compliance
Lightning Source LLC
Chambersburg PA
CBHW050053220326
41599CB00045B/7395